Organic Principles and Practices Handbook Series
A Project of the Northeast Organic Farming Association

Whole-Farm Planning

*Ecological Imperatives,
Personal Values, and Economics*

Revised

ELIZABETH HENDERSON
AND KARL NORTH

Illustrated by Jocelyn Langer

CHELSEA GREEN PUBLISHING
WHITE RIVER JUNCTION, VERMONT

Editorial Coordinator: Makenna Goodman
Project Manager: Bill Bokermann
Copy Editor: Cannon Labrie
Proofreader: Helen Walden
Indexer: Peggy Holloway
Designer: Peter Holm,
        Sterling Hill Productions

Printed in the United States of America
First Chelsea Green printing March, 2011
10 9 8 7 6 5 4 3 2 1   11 12 13 14

Our Commitment to Green Publishing

Chelsea Green sees publishing as a tool for cultural change and ecological stewardship. We strive to align our book manufacturing practices with our editorial mission and to reduce the impact of our business enterprise in the environment. We print our books and catalogs on chlorine-free recycled paper, using vegetable-based inks whenever possible. This book may cost slightly more because we use recycled paper, and we hope you'll agree that it's worth it. Chelsea Green is a member of the Green Press Initiative (www.greenpressinitiative.org), a nonprofit coalition of publishers, manufacturers, and authors working to protect the world's endangered forests and conserve natural resources. *Whole-Farm Planning* was printed on Joy White, a 30-percent postconsumer recycled paper supplied by Thomson-Shore.

Library of Congress Cataloging-in-Publication Data
Henderson, Elizabeth, 1943-
  Whole-farm planning : ecological imperatives, personal values, and economics / Elizabeth Henderson and Karl North ; illustrated by Jocelyn Langer.
    p. cm. -- (Organic principles and practices handbook series)
  "A project of the Northeast Organic Farming Association."
  Originally published: Barre, MA : NOFA Interstate Council, c2004.
  Includes index.
  ISBN 978-1-60358-355-8
 1. Organic farming. 2. Farm management. I. North, Karl. II. Northeast Organic Farming
Association. III. Title. IV. Series: Organic principles and practices handbook series.

 S605.5.H46 2011
 631.5'84068--dc22

                            2010049585

Chelsea Green Publishing Company
Post Office Box 428
White River Junction, VT 05001
(802) 295-6300
www.chelseagreen.com

Best Practices for Farmers and Gardeners

The NOFA handbook series is designed to give a comprehensive view of key farming practices from the organic perspective. The content is geared to serious farmers, gardeners, and homesteaders and those looking to make the transition to organic practices.

Many readers may have arrived at their own best methods to suit their situations of place and pocketbook. These handbooks may help practitioners review and reconsider their concepts and practices in light of holistic biological realities, classic works, and recent research.

Organic agriculture has deep roots and a complex paradigm that stands in bold contrast to the industrialized conventional agriculture that is dominant today. It's critical that organic farming get a fair hearing in the public arena—and that farmers have access not only to the real dirt on organic methods and practices but also to the concepts behind them.

About This Series

The Northeast Organic Farming Association (NOFA) is one of the oldest organic agriculture organizations in the country, dedicated to organic food production and a safer, healthier environment. NOFA has independent chapters in Connecticut, Massachusetts, New Hampshire, New Jersey, New York, Rhode Island, and Vermont.

This handbook series began with a gift to NOFA/Mass and continues under the NOFA Interstate Council with support from NOFA/Mass and a generous grant from Sustainable Agriculture Research and Education (SARE). The project has utilized the expertise of NOFA members and other organic farmers and educators in the Northeast as writers and reviewers. Help also came from the Pennsylvania Association for Sustainable Agriculture and from the Maine Organic Farmers and Gardeners Association.

Jocelyn Langer illustrated the series, and Jonathan von Ranson edited it and coordinated the project. The Manuals Project Committee included Bill Duesing, Steve Gilman, Elizabeth Henderson, Julie Rawson, and Jonathan von Ranson. The committee thanks SARE and the wonderful farmers and educators whose willing commitment it represents.

The first commandment of the Earth is: enough. Just so much and no more. Just so much soil. Just so much water. Just so much sunshine. Everything born of the earth grows to its appropriate size and then stops. The planet does not get bigger, it gets better. Its creatures learn, mature, diversify, evolve, create amazing beauty and novelty and complexity, but live within absolute limits.

—Donella Meadows, *The Laws of the Earth and the Laws of Economics*

CONTENTS

Introduction

Particularity and separability are infirmities
of the mind, not characteristics of the universe.
—Dee Hock, *Birth of the Chaordic Age*

Do you have trouble making decisions about your farm? Have you made decisions that seemed exciting at first but later turned out to be mistakes? Do you want to shape your farming and your life so that the many pieces fit together in harmony? If so, we believe this manual will help you.

Many of us who are drawn to organic farming and gardening tend to see the world as patterns of intricately interrelated systems. We share this perspective with so-called primitive peoples around the world who live with this connectedness that our fragmented, industrialized society has hidden from "advanced," "modern" people. Either from our innate cast of mind or from finding ourselves on a spiritual path, we have trouble with reductive science and the division of knowledge into separate disciplines, each with its own turf, jargon, and boundaries. We sense instinctively that we cannot do one thing on a farm in isolation, that if we make a change, everything that is connected with that change will shift too. We do not have to make a "paradigm shift" to grasp holism and the notion that the whole is greater than the sum of its parts. Organic and biodynamic teachings encourage us to think in terms of cycles and interrelationships.

But even with these good instincts we sometimes make bad decisions because we are distracted, under pressure, fail to consult with all the people involved, or do not think through the possible ripple effects. Whole-farm planning offers a way to slow down a little, and balance and systematize our decision-making process so that we do not indulge ourselves in foolish or expensive tangents that weaken our farm and leave us discouraged. It's a tool that functions like a pilot's checklist. A little training and self-discipline are involved, but the improved results will make it worth the effort. The method we suggest should not be taken as a

**Whole-Farm Perspective.
The world functions as
biological wholes within
wholes—cells, individu-
als, families, species,
"species co-ops,"
biological communi-
ties, ecosystems. Our
societies and economies
fit here within the still
greater wholes of water-
sheds, the biosphere.
(See also "Wholes
within Wholes" figure
following.)**

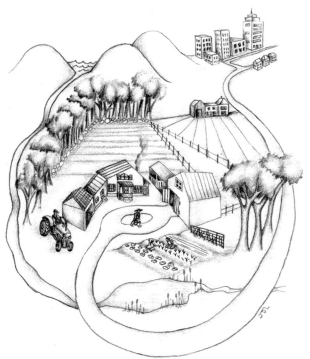

dogmatic recipe. Rather, it is a direction, which may lead some farmers, as it did the Flack Family in Vermont, to create their own system in concert with their particular style or spiritual or personal beliefs.

Why do whole-farm planning? What makes it more effective than other ways of managing farms? The answers to these questions lie in a quiet rediscovery through science that is fundamentally changing the way modern humans see and must manage the world.

The "new" perspective is to see that the world functions as biologi-cal wholes within wholes—cells, individuals, families, species, "species co-ops," biological communities, ecosystems. Our societies and econo-mies fit here within the still greater wholes of watersheds, the biosphere, the solar system.

Well, we knew that, at least in a general way—so what's so revolu-tionary? Scientists believe they are discovering the connective tissue that binds these wholes and their internal components. And they're finding the interdependence is far greater than they had realized. Bacteria, it

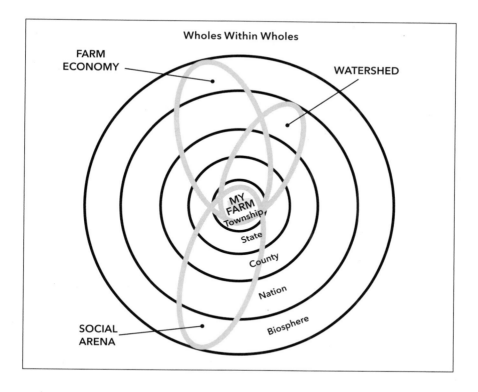

turns out, are the invisible, but very physical, tie that binds. The microbial realm locks together even many parts of our world that appear quite separate. The reasons for this interdependency have become increasingly apparent as our knowledge of the natural history of the planet has grown, and we've learned about:

- The common ancestry of all living things, rooted in the bacteria and other microbes that were the sole inhabitants on the planet for most of its natural history.
- The preponderant role of these microbes long ago in creating features of the planet we take for granted today (e.g., soil, oxygen in the air, clouds, and so on).
- The continuing central role these unseen actors play to maintain the delicate chemical and biological balance that keeps all the wholes on our planet in the dynamic equilibrium known popularly as health. Our own bodies contain upwards of 100 trillion

noninfectious bacteria, all doing an incredible variety of jobs that keep us healthy. Biologists say bacteria run (some say rule) the planet.

- That all the other species familiar to us today that evolved from microbes actually co-evolved. That is, they spent millions of years developing genetically together, and in the process learning to get along well enough that many became essential to each other's health and survival. The cooperative arrangements that formed and tested themselves for ages are the wholes that make up our world. Scientists who study these wholes know them as complex, self-organizing adaptive systems.

- That if we could quantify these free services, we would be staggered at their value in dollars. And that their value to us would increase greatly if, instead of damaging these systems, we learn to manage for their health and to better utilize this "natural capital."

What difference does all this make to how we plan and manage farms? Much of the knowledge and technology we employ in farming was and still is developed by specialists trained to focus so narrowly on a problem that they ignore the health of the larger wholes. Consequently, solutions that work in the short run often create even bigger problems in the long run. An example familiar to many farmers, both organic and conventional, is artificial fertilization via soluble salt chemistry. Along with the intended outcome of higher yields, farmers have discovered many unintended outcomes: soil compaction, nitrate toxicity in feed, weakened crop resistance to pests, nutritional loss in food, damage to the health of the soil food web, dependency on finite supplies of cheap energy, and pollution downstream. As students of the new paradigm say: When you make one change in a complex adaptive system (what in this manual we call a "whole") you never get only one outcome; you get many.

Does this mean that specialized knowledge is no longer useful? On the contrary, specialist-science has developed a powerful microscopic method for understanding the world by reducing it to its many details and pieces. But this reductionist science needs balancing and completion with a new, macroscopic way of doing science that recognizes that managing wholes is essential to sustainability. The goal of this whole-farm planning manual is to reintroduce just such a macroscopic method of making and testing decisions on the farm and in the larger wholes in which we live. Once learned, this decision-making framework will be something farmers can use to evaluate all the knowledge available from conventional agricultural science in terms of its impact on our farms as wholes. With enough practice it will become simply the way we automatically think and live in a world that functions in wholes. A commitment to whole-farm planning is, in a way, a growing-up, a Declaration of *Inter*dependence.

How Wholes Work: Lessons from Systems Science

Systems scientists are among the leaders at integrating conventional science and holistic thinking. Here is what they'd say good farmers need to understand about the systems they manage:

- The parts of the complex biological systems we live and work in are held together in trading relationships. Ecologists call these trades "metabolic exchanges."
- The nature, structure, and health of essential trading relationships that link parts of wholes are crucial to the health of the whole. Pollination is an example of a critical biological exchange; a fair contract between a farmer and a buyer is an example of a social exchange.
- Tracking system behavior over time is essential. Attention to long-term trends in our wholes is important in everything from soil organic matter to social relations. Change is one of the few constants of life and of complex systems. The equilibrium of these systems depends on how quick—and measured—is the response. Instead of holding to the rut of mythical permanence, farmers need to develop a ready sensitivity to change, as well as good balance in order to avoid "oversteer" and adding to the problem.
- More is rarely better for long. Growth in any essential part eventually becomes pathologic for the whole—true at every level of

> **Dynamic**, marked by continuous productive activity or change.
> **Equilibrium**, a state of adjustment between opposing or divergent influences or elements. **Resilience**, ability to recover from or adjust easily to misfortune or change
> — *Webster's New International Dictionary*

wholes: *E. coli* too numerous in our gut, cell growth that becomes malignant, more and more acres of a single crop that swells into a monoculture more prone to attack by insects and disease, farms so large the farmer loses contact with the land and needs satellites and computers to manage it, market share and merger increasing to unstable market monopoly. Continual growth eventually warps trading relationships and outstrips the resource capacity of any whole. System health depends on management for resilience and dynamic equilibrium, not endless growth.

Whole-Farm Planning and Sustainable Solutions

Taking the long view leads naturally to concerns with sustainability—what William McDonough calls "loving the children." A nationally known planner of architectural, urban, and industrial designs, McDonough was once invited to speak to the leading citizens of Chattanooga, a city at that time rated the most air-polluted in the nation. Having advised them at length on possible remedies for their city's ill health, he asked for questions. A hand shot up in the back, and its owner said, "Mr. McDonough, this sustainability stuff is all well and good, but just how long is it going to take?" Bill answered, "Forever, sir, that's the point."

In this manual, *sustainability* is taken to refer to three things, in order of priority. The first priority is sustainable management of the ecosystem(s) we aim to farm, because we now know that there are ecological imperatives that we must not compromise if agriculture and even civilization are to survive in the long run. Mother Nature does bat last! Second, within the constraints set by ecology, sustainability must apply to social relationships, first among the farm's decision makers, then moving outward with all those in the larger wholes that are important to the farm's health. All too often it is the state of social relations that holds farms from reaching their biological and economic potential. Finally, of course farms must be economically sustainable, but without compromising ecological and social prerequisites.

How is economic sustainability possible? To be realistic, we must admit that our present farm economy—a larger, global whole that farms nest in—limits progress toward ecological and social sustainability by rewarding most lavishly the least sustainable practices. Most farmers are driven to damaging industrial practices not out of disregard or selfishness, but because the market system leaves them little choice. We need to use the power of holistic planning to design a socioeconomic system that offers incentives and rewards for sustainable practices. The decision-making tool offered in this manual can be of help there too, in providing a method for stakeholders in our society to collaborate toward better design and management of that larger whole.

Despite the constraints just mentioned, whole-farm planning can put individual farms on the road to all three modes of sustainability since it is a multidimensional approach geared to the perspective of the holistic revolution defined above. The goal of whole-farm planning is to provide a way to tie all of the parts of a plan (environmental, social, economic) together in an integrated whole, for greater benefit in all dimensions.

What Kind of Tool: A Preliminary Sketch

A number of "whole-farm" planning tools exist whose Achilles heel is their narrow focus. Several are designed mainly to ensure environmental compliance, but unless considerable subsidies are involved, they remain economically unattractive to farmers. Others focus on technology transfer to raise productivity, but ignore profitability. Still others concentrate on short-run profit, ignoring the long-run health of the farm ecosystem. The organic farm plan required for certification provides a solid beginning for an assessment of the physical and environmental aspects of a farm and takes a bow toward the financial aspects by asking for a listing of markets. The USDA National Organic Program (NOP), however, insists that the social aspects of farming are "outside the purview" of its standards. The LLC certification application put out by NOFA-NY includes an optional essay question that allows a farmer to write an "Overview of Your Whole-Farm Philosophy," but this subject is definitely beyond the NOP. Most certifiers stick with ecological production.

By contrast, the planning tool we offer in this manual is immediately attractive to farmers because it starts by challenging us to articulate our quality-of-life expectations. This visioning process is powerful in several ways:

1. It brings out the ongoing essential connection between what we farmers most value in life and the kinds of activity or production we choose to pursue on our farms.
2. It stresses the equally strong link between the chosen kinds of activity and the health of the farm's resource base as it must be safeguarded far into the future in order to sustain those activities.
3. A whole-farm plan is a written, constantly evolving, working document that becomes the touchstone for a decision-making process; this ensures that all management decisions are accountable to the farmer's own quality-of-life desires and expectations. Equally important, when we are clear on our own goals, we can easily tell the difference between them and the goals of outside interests: agencies, financial institutions, research institutions, and purveyors of technologies and other farm inputs.

When it's time to solve a given problem, whole-farm planning helps us compare the tools or management options we're considering. In addition to testing against the goal statement, managers subject potential solutions to a series of other tests. Some resemble conventional business management, like comparing marginal return or ability to strengthen the financial weak link. Others test how well a decision addresses the long-term ecosystem needs for sustainable operation, asking questions such as: Are the sources of energy and other natural resources constant or finite? Are these sources benign or potentially damaging by their use? Does the decision make the most of the special opportunity farmers have to convert solar energy via photosynthesis into dollars? Evaluation of ecological impact focuses on four ecosystem processes in particular: the water cycle, the mineral cycle, energy flow, and changes in the biological community with respect to biodiversity. A last test considers the social and cultural dimensions of a decision. Underlying the testing strategy is the assumption that all management decisions are multidimensional in their effect, and that we must cover all the bases, every time.

This planning tool has helped farmers see the unique opportunities we have to use certain sustainable practices to our economic advantage in a farm economy where many sustainable practices are uncompetitive and unprofitable. It does this by helping us discover ways in which we are well positioned to utilize *natural capital* for economic gain while simultaneously conserving this essential capital. This tool does two things: it highlights the special advantage farming enjoys compared to other enterprises in its ability to capture solar gain via plant photosynthesis, and recapture that energy again and again across the food chain of the farm; and it focuses our attention on the synergies that natural history has created among plants, animals, microbial life, and humans in a farm's resource

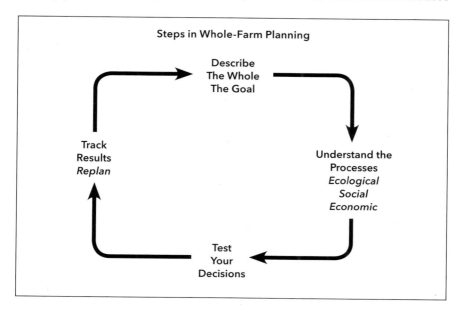

base and on how to capture their full potential. Thus we reduce outside inputs and make small farms more economically viable. When we do this successfully, our farms achieve their full biodiversity potential as well as the economic benefits that flow from harmony of function and form.

The deepest lesson of the latest scientific thinking—that our world is organized in complex nested systems—is that we must all begin to manage our affairs in a more holistic fashion, or else continue to court hidden losses and long-term failure. Farmers who invest in whole-farm planning are early adopters of this new perspective on management, and will be the early reapers of its advantages.

For Whom Is This Manual Written?

This manual is designed by farmers for farmers in the northeast United States and similar environments. We intend it for practical use, not passive reading. Only through constant practice does the new approach to decision making outlined in its pages become a powerful tool in a farmer's hands. Many farmers are already using whole-farm planning to improve the quality of their lives. Most of us who are using it successfully have formed groups to collaborate in the learning. With or without a facilitator, these learning communities provide inspiration and incentive, as well as co-learners to help each other along. We hope that this manual will make whole-farm planning accessible to more farmers, as well as homesteaders, gardeners, and rural communities, with the aim of strengthening our local food systems and contributing to a future of sustainable and peaceful communities.

two

Assessing the Whole: What Are We Managing?

Before we can start to make plans or decisions, we need to understand what we are managing. The answer may seem simple—well, it's the farm. But when you think about it carefully, you begin to realize that the borders reach farther than is obvious at first glance. When we say "whole farm," we mean *all* of the human, physical, and financial resources that make each farm a unique social and ecological entity.

To create the fullest inventory, it helps to think about three elements of a farm—people, assets (both physical and mental), and money, and to put your observations down on paper as the groundwork for your plan.

People

First of all, consider the people. Let's draw a series of concentric circles. In the center are the people who are most directly involved in the operation, the ones who will be making decisions and living with the consequences: the farmers, their immediate families, and any close partners, long-term employees, landlords, or lenders who might veto your plans. These are the people you will want to meet with when you draw up goals and plans for the farm. In the next circle are those who are indirectly involved with the farm: customers or CSA members, interns or apprentices, family members who do not work or live on the farm but might visit, support, or inherit it. In the outer circle are those who are affected more peripherally: neighbors, other organic farmers, the local buying public, and even more distantly, all the rest of sustainable agriculture, agriculture, and the food system as a whole.

For co-author Elizabeth Henderson's Peacework Organic Farm, the people in the center would be Elizabeth, her partners Greg and Ammie,

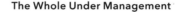

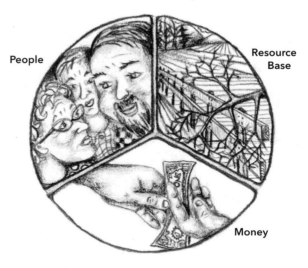

their daughter Helen, and Rebecca Kraai, from whom they rent land, equipment, and buildings and trade work back and forth. In the next circle would be their farm interns, the members of their CSA core group, the other CSA members, Elizabeth's son, daughter-in-law, and father-in-law, and Greg and Ammie's parents. The outer circle would include their neighbors in Arcadia, the Abundance Cooperative Market, Lori's Natural Foods, the other farms in their cooperative intern-training project, and other area organic farmers. Most peripherally involved would be other county farmers and colleagues in sustainable agriculture in our region and around the world.

Physical and Mental Assets

Next are your physical and mental assets. Draw another set of concentric rings. In the center is your farm and the level of skills and knowledge you bring to your farming; the number of acres of arable land, pasture, brush, forest, and so on; the kinds of soils; the plants and animals; the climate, buildings, equipment, and tools; the available library or Internet access; and the location of the farm in relation to markets and transportation.

Among your assets you should count the natural capital* available—"all the goods and services provided by nature that contribute to the well being of humans and every other species on the planet." In the next circle are the neighboring pieces of land, other land you might own or rent, farmer groups or farming organizations you belong to, the extension service and university resources. And in the outer circle are the county, the region, and connections around the world.

If you certify your farm as organic, the paperwork you have already done will be helpful in compiling information on the physical aspects of your farm.

For Peacework, the physical assets are the eighteen acres of well-drained silty and sandy loams on which we grow vegetables and that we share with woodchucks, deer, raccoons, mice, moles, many birds and other wildlife; the treelines surrounding the fields; a large barn for storage with an attached greenhouse; a smaller barn used as a packing shed; a cold frame; a hoop house; three tractors; an assortment of implements, hand tools, and farm supplies; a creek, a smaller stream, a spring that runs steadily all year round, and a well. We enjoy (or suffer) the climate of the lake-effect zone of western New York, which is a Zone 5 growing area. The land is on a back road, five miles from the nearest village and an hour's drive from the nearest city. Our human assets include Elizabeth's twenty-one years of farming and many more as an organizer and writer, Greg's eleven years of farming and homesteading, and Ammie's three years of farming and thirty of gardening as well as her studies in soil science and experience as an organic farm inspector. We have an extensive library of farming books and periodicals, Internet access, and close connections with some researchers. Surrounding our fields are the hayfields and woods of Crowfield Farm, with only one field bordering on a conventionally-managed neighboring farm. We are active members of NOFA-NY, read the newsletters and participate in field days, farm tours, workshops, and conferences. Recently, the Cooperative Extension and Cornell researchers have taken an interest in organic farming. As NOFA's representative to

---

* Natural capital includes both mineral and biological raw materials, renewable (solar and tidal) energy and fossil fuels, waste assimilation capacity, and vital life-support functions (such as global climate regulation) provided by well-functioning ecosystems. From Visions for a Sustainable Future (http://www.sustainablescale.org/AttractiveSolutions/VisionsForASustainableFuture.aspx)

International Federation of Organic Agriculture Movements (IFOAM), Elizabeth is in contact with organic farmers and farming organizations worldwide.

To assess the physical assets of the farm, mapping is extremely helpful. Make a map of the land, buildings, and neighboring properties. Include existing vegetation, ponds, streams, and wetlands. Obtain topographical and soil maps from the soil and water conservation district so that you have complete information about the characteristics of your soils and layout of the fields. Learn all that you can about prevailing weather and winds. If you have livestock, do a census of your animals, including their age, sex, and state of health.

Money

Finally comes money, or financial resources. In the center of your rings are your current assets, the money or barter directly available to you from farm or off-farm earnings, savings, or other sources, and your ability to manage them. In the next circle are family holdings, existing cooperatives with which you might work, financial aid available from government or banks, potential loans and grants, and any investors or supporters of your farm. And farthest from your direct control is the broader farm economy in the region, the nation, and the world.

Peacework's financial resources center on the investment in the farm. Elizabeth fronted the initial money and equipment, and her partners are gradually building their equity so that the shares will be equal. They supply produce to a CSA of 280 families who commit to a yearly contract with them, and make contributions to the farm capital fund. As a result, the farm has no debt or need to borrow money. The local bank agreed to give Elizabeth a mortgage on a house partly on the strength of the CSA. Over five years at this farm, Peacework has developed elaborate connections for barter and mutual help with its landlords and neighbors. With Greg's background in political economy and Elizabeth's attempts to understand the food system, we keep a close eye on the national and international economy.

For yourself, the more complete a picture you can make of where you are starting from, the more realistic and creative your plans and decisions are likely to be. If you are the only person directly involved in managing and operating the farm, your process is relatively simple. Most farms, however, involve the farmer's family, whether willingly or not. Family members supply labor and financial support. The long-term health of your farm may be in danger if you do not communicate well within the family and include everyone in establishing goals for your farm. Family dysfunctions and divorces have weakened or destroyed too many farms. Any plan for the whole farm has to begin with all of the people whose lives revolve around the farming.

To be clear about your farm's finances, it is essential to develop a business plan. (Banks and other lenders usually require a business plan before they will lend you money.) There are many resources available to help you do this. In many counties, the Cooperative Extension offers trainings. In most of the Northeastern states, an organization called NxLevel® offers courses. The NxLevel motto is "Helping Entrepreneurs Reach the Next Level of Success." In agriculture, their ten-session course, "Tilling the Soil of Opportunity," aims at "individuals who have started or are thinking about starting an agriculture-based venture that is not tied to large-scale, commodity-style production." (See their Web site: www.nxlevel. org.) Although the Web site does not mention clarity about personal goals, Kathryn Hayes, who teaches these courses in Massachusetts, says that she always starts with an exploration of students' quality-of-life goals. The Land Link programs in the northeastern states also provide workshops, such as "Using Financial Tools for Farm Business Success!" and "Transferring the Farm." As Rhonda Janke, formerly at the Rodale Research Institute and presently Professor of Agronomy at the University of Kansas, points out, "Economic sustainability means not just generating a positive cash flow picture, but doing so without draining a farm's equity or net worth and by taking into account depreciation and replacement of farm assets such as buildings and livestock breeding stock."

Do not limit your thinking to conventional sources of funding. If your farm offers important services or resources to the community, you may have social capital available. Social capital, as Janke points out, includes: "one's reputation in the community, extended family relationships, positive working relationships with landlords, local mechanics and other businesses . . . that can translate into savings of dollars, and access to resources that would not otherwise be available." The members of a CSA and regular customers at farmers' markets or farm stands will sometimes lend or give money to a farm to help make it more sustainable. They might purchase produce in advance. There may be neighbors or other farmers who are willing to barter goods or services. If you like working with people, there are all kinds of groups that will come to do work projects at your farm.

Tallying all of these internal and external assets and resources will give you a complete image of your whole farm. A realistic appraisal of your baseline provides you with a solid basis for planning for the future and monitoring the progress of whatever changes you decide to make.

three

Goal Setting Is an Ongoing Process

The successful application of holistic management depends
entirely on your ability to think and be creative, and this in turn
depends chiefly on your mental, emotional, and physical health.
—Allan Savory, *Holistic Resource Management*

Yogi Berra is said to have remarked, "If you don't know where you're
going, you might end up somewhere else." This seems obvious, yet farms
rarely have a goal that all the decision makers and main participants understand clearly and agree to. A written, working document, created in a
goal-setting process in which all the people central to the farm participate,
is a crucial tool for effective planning, monitoring, and management. This
set of goals can serve as the touchstone by which we as farmers judge
all important decisions, helping to balance improvements in our overall
quality of life—including financial security and social and community
relations—with the health of the natural ecosystem within which we
work. Returning regularly to the goal document as you make decisions
will suggest changes or more detail that you want to add to better express
what you want. A cyclical planning process clarifies what qualities you
really desire in your farming life. Thinking about your goals as an evolving document also makes it less scary to start writing them down—you
can change them, you will *want* to change them, so don't worry that your
first version is not "perfect."

In order to integrate the many aspects of your farm, it helps to write
your long-term goal in three parts: Quality of Life, Forms of Production,
and Future Resources Base. Everyone in the farm's inner circle—the
primary farm operators, close family members, and nonfamily members
who have a financial stake in the farm—should participate in the process
of working out this three-part goal. Keep in mind that you should stress

in your goal what you *want*, not what you do not want. Figuring out how you will get there is a separate process. The group of people central to your farm will most likely have disagreements about some of the particular steps you will take. But if you can reach agreement on the overall long-term goals, it will make choices easier, and the process of goal creation can strengthen your group.

The entire central farm group may want to work out values together in a cooperative brainstorming process. The object is to discover shared values—and not all values are shared. Negotiating goals that include all the members of your farm team requires time and a willingness to listen closely and respectfully to one another. Your process might take the form of a series of meetings where you go over each section word by word and make revisions. More often, this task is made easier if each member does this separately and then the group compiles the goals for the entire farm by combining these statements. If there are differences in age, experience, or assertiveness among the members of the group, this structure makes it more likely that you will give full value to each individual. Have each person compile his or her own version of the goals, allow everyone in the group time to digest these statements, and then come together and choose the pieces from each version that you prefer. Experience shows that, gradually, you can come to consensus with a draft that everyone likes. For these goals to be an effective tool, everyone in the group must embrace them with enthusiasm.

Goals Part 1: Quality of Life

This section of your three-part goal sets out your most deeply held and cherished personal and spiritual values, and the quality of life you wish to lead. There are some helpful visual approaches to getting at these values.

First, you can "mindmap" what you value: draw larger circles for what's more important, draw lines connecting things you see as related.

Second, you can think about what you would like people to say about you at a memorial service celebrating your life. (Ed Martsolf, who gives trainings in whole-farm planning, uses this exercise derived from the work of Steven Covey.) What are the outstanding qualities and accomplishments by which you would like to be remembered?

At a Martsolf training Elizabeth attended in 1998, she came up with this quality-of-life goal: "fresh air, close to nature, working closely with other people, independent (no debt—political, financial, or psychological), with time for my family, culture, politics, learning and spiritual growth; human relations on the basis of mutual respect."

Third, answer the following questions in writing. Include anything that is important to you. (John Gerber, in *The Holisticgoal*, is the source for this exercise. He starts you off with a series of questions "designed to get your mind engaged in the process." Please see Gerber's Web site, www.umass.edu/umext/jgerber/hmpage/hmpage2/wordlist.htm for additional worksheets to help you with this process.)

1. What are the three things that you really like about your life right now?
2. What three things about your life would you like to change if you could?
3. Name several talents or skills that you have and that you enjoy contributing to your farm, family, and community.
4. What are three talents or skills that your spouse (or parent, child, hired person, or business partner) contributes to the farm, family, and community that you really appreciate.
5. What would you like to be doing ten years from now? What sort of changes would be necessary to bring that life to reality?
6. What was the biggest change on your farm in the last five years? How did that change affect you? How did it affect your spouse, child, hired person, business partner, or neighbors?
7. What is the best part of your day?

Once you have some writing done, take phrases from your QOL statement and list them in the left-hand column of a table like this one:

Quality-of-Life Phrase	What Will That Give Me? **What improvements will this provide in my life?**
1.	
2.	
3.	
4.	

Now, for each phrase, go ahead and examine what that would give or provide you. Ask again and again, changing the phrase as necessary, until you are sure it expresses a value that's truly fundamental to your desired quality of life. This is the beginning of a really fundamental QOL statement.

Once you have ten or twenty phrases that seem truly fundamental, you can rewrite your quality-of-life statement more clearly. Start with "I want . . ."

When each decision maker has written a quality-of-life statement, you should work together to create a combined statement. You may want to go back and use these tables again for the combined statement. Don't be too hard on yourselves. Get a draft statement written to use for making decisions. You will probably change it later.

If you are still stuck on the quality-of-life statement, on the following page is a list of words you might use to help you get started writing. First, cross off the words that you don't want in your QOL statement. Next, circle the ones that resonate with you and you may want to include. Finally, try to cut the list to ten or twelve that will be used to begin writing your QOL statement. Once you have ten or so key words, start writing a QOL statement using them.

For some people, a visualization process such as this one may work better: Close your eyes and breathe deeply. Relax and imagine a place where "everything is perfect." Take a few minutes to imagine this place and then begin to notice . . .

Key Words for Your QOL Statement

debt free

comfortable

freedom/opportunity

health

fun

financial security

spiritual fulfillment

community
  connection

challenge

personal growth

contribution

recognition

close family

open communication

connection with
  neighbors

family

work together

time for rest

meaningful work

learning

companionship

welcoming home

excited about work

excited about play

live simply

content

secure

serene

happy

joyous

satisfied

good relationship

productive

stable

without restrictions

wealthy

successful

ethical

change

famous

helping others

honest

independent

adventure

privacy

service

religion

reputation

sophistication

status

wisdom

order

tranquility

loyalty

integrity

physical challenge

pleasure

affection

creativity

expertise

fast-paced

excitement

cooperation

knowledge

leadership

inner harmony

nationalistic

compassion

truth

patriotic

idealistic

laughter

open

promote

fun environment

options

breathe

grounded

redefined success

soul

eat well

community
  self-reliance

live free

connected

love

power

respect

ecological balance

diversity

culture

ambition

the people there—what are they doing?
the immediate surroundings
any plants and animals
the further surroundings—landscape and horizon
the colors
the smells
what else do you see?

Now, working from this place where "everything is perfect," begin writing. Start with the things you "truly desire." It's okay to write, "I want (fill in the blank)." Or use the word list to get started if that helps. Circle those things you value on the list and then begin writing "I want" statements using these terms.

As an example for inspiration, here is Karl North's QOL statement for Northland Sheep Dairy:

1. Economic and physical security
2. Affectionate, collaborative, stimulating, trusting relationships
3. Fun and challenging
4. Political freedom
5. Environmental health
6. Egalitarian social justice
7. Cultural stimulation
8. Self-respect
9. Life-long learning
10. Beauty and human scale

Goals Part 2: Forms of Production

To achieve and sustain the quality of life you desire, you need to produce or create things. These products will take different forms to satisfy different values you described in your desired quality of life. Therefore, some forms of production you will want may be economic, but others may seek to satisfy social, cultural, spiritual, or political values that you have identified as important to you.

To describe the forms of production you want, think about what farm enterprise(s) you might operate to enable you and your family to afford to live on your farm. How much of your life energy do you want to put into the farming? Full time? Part time? How much economic security do you need? If you listed in your quality-of-life statement time with family and friends, how will you make the time in your schedule for this? Is there off-farm work that any of the members of the farm team want to do? If, like the Norths, you want "political freedom" and "egalitarian social justice," what concrete steps will you take to make that possible? As Gerber puts it, "What are we doing or not doing now, or what don't we have now, that is preventing us from achieving this aspect of the quality of life we want?"

Here is a worksheet-style approach to defining your production goals. List each of your quality-of-life statements in this table, and then answer the question, What do I have to produce to get this?

Quality-of-Life Phrase	What do I have to produce to get this?
1.	
2.	
3.	
4.	

Next, ask the question, What personal, social, and cultural—as well as economic—forms of production do I want or need?

Then ask the following questions:

- What do I know how to do?
- What do I want to do? What is my dream?
- What will my resource base sustain?
- What will the market bear?

Then write down what forms of production will offer each of the things you named in the first part as valuing.

Here is the first draft of Elizabeth's production goals, written before Peacework Farm was started:

> Organically managed market farm as self-sufficient as possible, with regional links of interdependence with other farms, producing year-round supply of the most-alive-possible fresh and stored vegetables for 200 families with maximum participation. Excellent quality, sense of order, and calm. Optimal mechanization. Within ten years, replace myself completely in the role of farmer-organizer.

The forms of production of Karl's Northland Sheep Dairy are more developed and detailed:

1. Profit and enjoyment from dairy sheep and other livestock, other farm products.
2. Homestead production for a healthy degree of family self-sufficiency.
3. Farming for collaboration, innovation and education: sharing the work, and the knowledge gained.
4. Work that balances the physical and intellectual, business and friendship, seasonal integration, stewardship and profit.
5. Marketing to enhance the local food economy.
6. Development that achieves an attractive harmony of function and form.
7. Civic and other political work toward an environmentally healthy, culturally creative, democratic/egalitarian civilization.
8. All forms of production will obey the following ecological imperatives, which are principles of management intended to favor sustainability, or permanent ecosystem health:
   - Use holistic, site-specific designs
   - Capture interspecies synergies

- Use local, current solar gain
- Respect nature's cycles: waste = food, etc.
- Design to appropriate scale

(A useful tool for these and other issues in financial planning is Rebecca Bosch's *The Organic Farmer's Guide to Marketing and Community Relations*, NOFA Handbook Series).

Goals Part 3: Future Resource Base

Consider the social, economic, and physical environment of your farm: What must it be far into the future for your production and quality-of-life goals to become possible, or to be maintained once they are achieved? What soil quality is necessary? How much biodiversity? What level of cleanliness of the air, water, and soils? What changes must occur in your neighborhood, town, county, state, region, and even the entire world? As Gerber puts it:

Make sure you include the following three components of the Future Resource Base:

1. Yourself. What must you be like in the future? Describe yourself as you hope others might describe you.
2. The landscape. What must the land be like in the future to support the quality of life you desire? Describe the physical landscape in which you hope to live. Describe the natural resources either immediately under your management or in your region.
3. Your community. What must your community be like in the future to support the quality of life you desire? Describe the attributes of the community you hope to live in. Consider things like common values, civic activities, health services, schooling, recreation, and other community resources.

Elizabeth's first sketch read like this: "Land regenerated, healthy soil, biodiversity below- and aboveground, lots of wildlife, and not too many close neighbors."

After five years of work, the Norths and their farmmates produced this more developed "future landscape" statement:

1. Stable, living soils with 15 percent organic matter.
2. Optimum biodiversity appropriate to each type of land use: perennial forage, forest, hedgerow, orchard, gardens, eventual annual cropping for market, farmstead site.
3. A farm ecosystem that achieves a reasonable self-sufficiency of inputs, particularly for feeding the soil community, plants, animals, and people in the whole. This ecosystem will include:
   • Plant and animal genetics appropriate to this design.
   • Enough open space for an effective grass/ruminant complex (described elsewhere).
   • Hedgerows accessible in all paddocks.
   • Efficient cycling to ensure ecosystem health.
   • Forest and other biomass production and solar capture to meet fuel, other energy, and compost material needs of the system.
4. Enough open space (fields) for production needs.
5. A conservation easement or other agreement that legally protects the farm from nonfarming development.
6. Regional impact of the farm as a trend-setting model of sustainable design.
7. Younger managers taking over in time to maintain the farm.
8. A local community that remains largely rural, and becomes highly interactive and interdependent—economically, socially, and culturally.
9. A thriving regional economy, reasonably self-sufficient in food, energy, and shelter.
10. World peace via policies of economic security for all,

replacing economic imperialism and other hegemonies
and discriminations with democratic decision making.
11. A reputation with others that fosters progress toward the
other parts of the three-part goal.

After three years of farming together, Elizabeth and her partners,
Ammie Chickering and Greg Palmer, took their first drafts and came up
with this combined set of goals:

- To produce safe, nutritious, hand-crafted, fresh food for area
families.
- To reduce to the minimum the miles our food travels to reach
the tables where it is eaten.
- To minimize or eliminate the soil erosion that usually comes
with crop production.
- To maximize biodiversity above and below ground.
- To recycle nutrients and reduce our dependence on nonrenewable resources like diesel fuel and other off-farm inputs.
- To create an attractive and orderly farmstead that complements
the beauty of Crowfield Farm.
- To create a work atmosphere that is calm, cheerful, and welcoming; that allows us to enjoy the rhythm of the seasons, and to be
attentive to the natural beauty around us.
- To create an environment that is safe, fun, and educational for
children, including our own.
- To involve the community in our farm, and our farm in the
community; to improve local food security, social justice, and
cooperation among farmers and between farmers and consumers.
- To continue to learn and to share with others the knowledge and
skills we acquire.
- To make a modest living for our two families: we are blue-collar
farmers; we enjoy physical labor and have no desire to become
managers or exploiters of other people's work.

four

Making Decisions That Consider All the Variables in Your Whole . . . and Their Interdependency

The only wealth sustaining nations in the long run is derived from biodiversity, yet no conventional economic text even mentions the four fundamental processes that sustain biodiversity: water cycle, mineral cycle, community dynamics (biological succession), and energy flow.
—Allan Savory, *Holistic Resource Management*

What You Need to Understand about the Ecosystems That Sustain the Farm

Ecosystem health, today and far into the future, is the foundation of farm production and all other activities that contribute to a farm family's quality of life. As you make decisions about your farm, you will want to evaluate them for their ecological impact, just as you measure them against the farm's three-part goal. This chapter will offer four windows into the system to help build a better understanding of the processes that drive it.

Of all the ecosystems that can affect farming or be affected by farming, the ones on the farm are those that the farmer can best manage. They will therefore be the main focus of this chapter. But as the farm ecosphere functions within the larger wholes of aquifer, watershed, climate zone, and planetary biosphere, we should keep in mind that boundaries really exist only in our heads, not in nature. Water, commodity-energy, many minerals, and all kinds of living things come onto the farm from somewhere else and eventually go somewhere else. Some of the effects of these flows are predictable and even manageable. We know what causes the polluted estuaries of the Susquehanna and the Hudson rivers, the rain that keeps adding acid to our soil, and even the ever-warming winters over this century. Some of the effects are not predictable, like the proverbial butterfly of chaos theory that by flapping its wings in Tokyo can cause a hurricane in Tampa. What can we know and manage regarding the ecosystems on our farms?

There are four fundamental ecosystem processes:

1. The water cycle
2. The mineral cycle
3. The dynamics of the biological community
4. The energy flow

These processes are all at work on your farm, either for or against each other depending on how you manage them. The weakest one will be the limiting factor that determines the health of the whole ecosystem. We discuss them separately here only to help you learn them as windows into the system.

The Water Cycle

Water comes onto the farm from the sky, from upstream, and from uphill across and through the ground. It leaves the farm by flowing downstream, down through the soil into the aquifer, and by evaporation. Since water drives the growth of all life and its shortage is a major limit to growth, farmers need to know how to capture it and hold it in appropriate quantities on the farm. In other words we need to recognize the difference between an effective (as measured against our three-part goal) and ineffective water cycle.

The two most important places farmers can capture water are in soil and plants. A farm with an effective water cycle will have:

- Soils high in organic matter that provide absorptive material, pathways for water to flow into them, and food for soil life.
- A dense community of soil organisms—themselves mostly water—that create water channels and break surface crusting.
- Complete cover that shades the soil and slows water transpiration.

Effective Water Cycle. Farmers need to work to capture and hold water effectively through tillage choices, planting practices, and preservation of soil with trees and bushes.

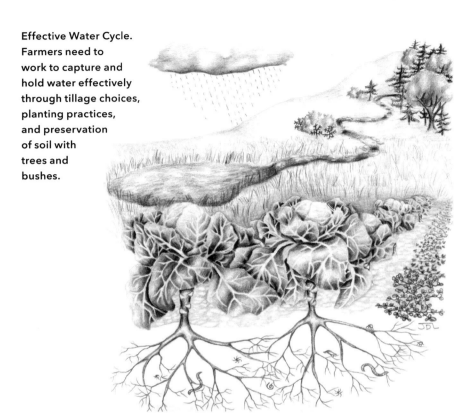

- Trees and bushes whose deep shading of the soil provide for more water storage than most other farm environments.
- Windbreaks to slow transpiration.

How can farmers achieve an effective water cycle? Here are the components of an effective water cycle:

- Following a strong farm program to build soil organic matter. Green and animal manures provide temporary fertility, but rarely contain enough carbonaceous material to build soil over the years. Getting your soil carbonated requires large carbon inputs, often much more than the farm can bootstrap with the depleted soil with which most farmers start. Imports from nearby off-farm sources will help to jumpstart the program. High-fiber biomass, such as mature hay, straw, or leaves, contains the most carbon, but to avoid temporary disruption of soil fertility, it needs composting before incorporation. The fastest way to build soil is by cycling part of the biomass (grass and hay) through herbivores and composting the rest of it with the manure the herbivores produce.

- Using hay and pasture in crop rotation, to build soil organic matter.
- Managing the grazing of grasslands for deep plant roots, pulsed root die-off that adds organic matter, and enough winter cover to catch snow. See chapter 5 for a detailed discussion of grazing management.
- Growing cover crops, themselves mostly water, between row crops and after crops are harvested.
- Covering soil with litter, crop residues, and mulch.
- Capturing water for irrigation with ponds and key-line plowing. (An introduction to the practice of key-line plowing is available online at www.soilandhealth.org/01aglibrary/010126yeomansII/010126toc.html.)

The signs of an ineffective water cycle include:

- Bare soil that produces runoff, especially when pounded by heavy rains.
- Soil compacted by chemical fertilizer, machinery, or excess tillage.
- Plants with poor root growth.
- Excessive runoff.
- Slow water penetration.
- Wells or springs drying up.
- Floods downstream.
- Droughts. Water shortages in towns and cities.

The Mineral Cycle

The plants and animals that farmers grow need a wide variety of mineral nutrients. Conventional farming methods neglect nature's way of providing nutrients through a cycling process. Artificially manufactured inputs,

Vegetation dies and decomposes

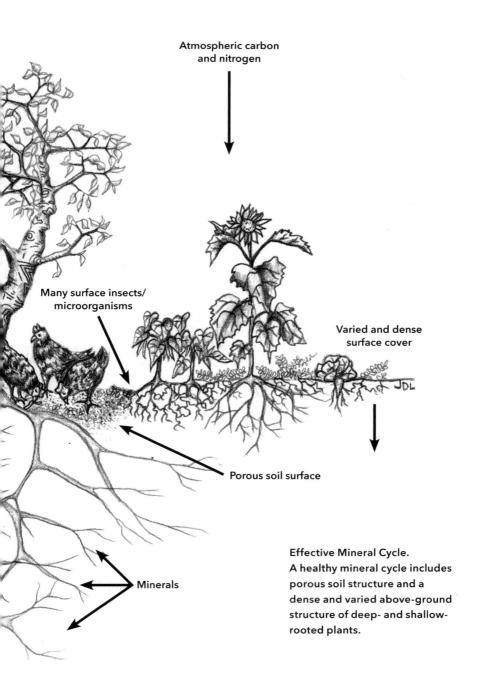

Atmospheric carbon
and nitrogen

Many surface insects/
microorganisms

Varied and dense
surface cover

Porous soil surface

Minerals

Effective Mineral Cycle.
A healthy mineral cycle includes
porous soil structure and a
dense and varied above-ground
structure of deep- and shallow-
rooted plants.

which are water soluble, are hard to control: plants get too little or too much. Chemical fertilizers often create a toxic environment for soil life. Under certain conditions, these chemicals bind to other chemicals in the soil and become unavailable to plants, or leach or wash away as runoff and pollute surface waterways and underground aquifers.

A well-managed natural mineral cycle provides nutrients that plants and animals need, and does it without creating new problems as it solves old ones. How does this cycle work?

The atmosphere provides carbon and nitrogen, and bedrock slowly turning into soil provides most of the other minerals. The vast, teeming microbial communities in and on the ground process minerals from the soil and air into forms plants can absorb. By feeding on plants and other animals, creatures carry minerals through the food chain. Along with insects and other decomposer species, microbial communities are essential to the decomposition of plants and animals, returning minerals to the soil.

At some points in the cycle, minerals take such forms that water can leach them deep into the soil. In well-watered climates like the northeastern United States, deep-rooted plants and trees perform a necessary function of pulling these minerals back into the cycle.

Here's how farmers can manage for a healthy mineral cycle:

- Provide environments at the soil surface and underground that are conducive to highly active biological communities: soil well fed with organic matter, sufficient surface cover for the decomposers, and a porous soil structure.
- Incorporate many different species into the aboveground farm system—a dense and varied cover of deep- and shallow-rooted plants, and animals whose gut can process all types of biomass.
- Bring the most minerals into the cycle—e.g., bring compost material to the farm—and keep them from escaping.
- Import concentrated minerals for balance only as a last resort. Often soil that seems to be lacking minerals actually needs better biological activity to make these nutrients available. If possible, import minerals in organic form—animal feed, for example— that can serve the dual purpose of nourishing your livestock while adding needed minerals.

The Dynamics of the Biological Community

The conventional approach to agriculture assumes that production species—crops and livestock—are more important than the other species in the whole, such as weeds or wildlife. But the organisms on a farm exist in interdependent relationships, as individuals do in human society or any other whole. These longstanding cooperative arrangements are far more important than the forces of competition to the health of the biological community. The survival of every species depends on turning other species into food, often with the help of intermediaries. Much of what people regard as competition actually performs an ingenious function of stabilizing population that works in the community's—and the crop's—favor.

The natural direction of development of these communities is roughly toward more complexity, an evolution through stages of succession where more biodiverse communities gradually replace simpler ones, gaining stability and resilience in the process. Agriculture, by definition, arrests succession to some degree, but farmers can benefit from the successional force of nature by:

- Creating habitats for nature to fill.
- Designing farms to become as rich in plant and animal species as production goals will allow.
- Strategically encouraging natural succession to change environments in favor of desirable species or to reduce undesirables.

The physical and climatic character of a region helps determine the biological communities as they pass through stages of succession. As farmers discover more about how succession works in their locale, they can partake of the opportunities nature provides to build stability and resilience into the biological systems on which their crops depend. These discoveries, in turn, can lead to greater productivity. With regard to perennial crops such as grasses or raspberries, for example, you might ask: What mix of forages will create a lasting community? What ground-cover plants and their associated microbial and insect life will keep my raspberries healthy over the long term? As you answer these sorts of questions, you can improve your description of a future landscape in your three-part goal so that it will bless the activities in your goal.

The Energy Flow

Living things depend on energy from the sun. Besides giving warmth, solar energy is initially captured and converted to usable life-currency by plant photosynthesis. It then cycles through the food chain as some form of carbon, as either living or dead tissue, or as atmospheric gas. Farmers have unique opportunities to profit from solar energy simply by the way they farm. Focused farm management can dramatically increase energy capture, and thus food production, not only for humans but also for all living creatures in the farm's ecosystem, thereby enhancing the health of the whole system. The sun is cheaper than any other form of energy on the market: it's free. Boosting photosynthesis is a farm's best capital investment.

In the simplest form, we can picture solar energy flows as two pyramids built on a green plant base, feeding the aboveground food chain and the underground food web. Each level of the pyramids is smaller than the one "below" to indicate that the food chain is hierarchical: each level must live

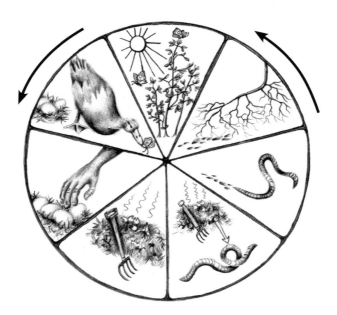

The Energy Cycle. Green plants capture solar energy, the real fuel of the farm. That green plant base, including broadleafs and legumes, is the source of the energy that drives the food chain that drives the farm.

off only the surplus of the ones below it—not the entire population—so each level supports less life.

The other ecosystem processes—the water and mineral cycles, and community dynamics—must be functioning well to maximize energy capture. But as a farmer, you can facilitate the sun's work in your operation in two ways. First, develop the green plant base to its maximum potential. It's the foundation for all the other solar energy activity on the farm. Pay special attention to three dimensions that influence the capacity of the plant base to capture energy:

1. The density of the plant base. Bare ground converts little sun to carbon. Close spacing of row crops, growing plants in beds, sowing cover crops between rows or beds, and hay and pasture managed to maintain good cover are all important here.
2. The leaf area of the plants. Broadleaf plants are generally better converters. Grasslands are most effective when they include broadleaf species.
3. The growing time and rate. Choosing varieties of cropping and grassland species to lengthen the growing season and keep something growing during extremes of temperature and humidity will maximize energy conversion.

Second, keep offering that energy (biological matter) to the next taker to get the most out of it before it escapes your whole. You can do this by enlisting the appetites in the upper levels of the food chain, animals, birds, insects, and decomposer organisms that will directly or indirectly benefit your farm system.

As you practice whole-farm planning you will ask how every decision affects the four ecosystem processes. You should also ask: Do these ecosystem changes move my farm toward my stated three-part goal? Refer to this chapter to help you think about such questions.

Many ecosystem changes are slow and hard to see without careful monitoring. Chapter 6 describes monitoring methods that are an important part of whole-farm planning.

What You Need to Understand about Social Structures and Processes

Both the production and ecosystem processes and the financial planning on your farm can be well organized, yet they can fail to move you toward your goal if the social relationships that are important to the farm are not working well.

On the mythical family farm, husbands and wives, parents and children, friends and neighbors all work together in cooperation and harmony. Mutual respect and appreciation govern relations between sexes and generations. When differences of opinion occur, the people involved take the time to sit down and negotiate a mutually acceptable resolution. If only we could live up to this myth! In reality, all too often tempers flare; in the pressure of the moment, people who care deeply for one another say angry or thoughtless things and feelings are hurt. How can we move our reality closer to the myth?

At the Northeast CSA (Community Supported Agriculture) Conference in 1999, a diverse group gathered for a workshop on the topic "Women and Men Working Together." There were two of us who had left farms where relations did not work out with partners, two young men who were about to start farming with their female partners, two young women who farm with men, a woman who has been farming for many years with her male partner, and a woman who is still farming with her ex-husband although they live separately. Everyone contributed to our lively discussion. We concluded by drawing up a series of recommendations to help ourselves and others work more harmoniously and productively together:

- Don't let irritations and disagreements accumulate till an explosion occurs. When a difference occurs, note the context and skip the generalizations. Use conflict resolution and hold regular meetings.
- Differences in sensibility can be a source of richness or of increasing irritation. Honor deeply your own perspective and the other person's.

- Balance expertise, and nurture leadership skills. Take charge of separate areas and exchange these when possible to gain perspective of the other role.
- Write a contract (Nolo Press publishers has helpful models. See www.nolo.com).
- Resolve issues by the end of the day—don't go to bed angry or hurt.
- Learn good teaching skills—how to give enough information and then the space and time to absorb it.
- Do not overburden a partnership by always having the more skilled partner teach the less skilled. Turn to others for instruction. Our community needs to offer training in critical skills.

Some of us take this farming way too seriously. Build in play and social time. Have fun!

People do not automatically develop the social skills necessary to carry out these recommendations. We often fear confrontations and avoid conflict. We tend to assume that we have social skills as part of our cultural heritage, but rugged individualism rather than cooperative values and customs has been the strong suit in our culture. In this section of the manual we share some of the social processes we have found helpful to develop skill in this area.

One of the biggest obstacles to effective, conflict-free relationships is the assumptions we make about each other, the actions of others, and the beliefs others hold about us and about the world. Each individual looks at the same situation with different assumptions, makes different inferences, and therefore draws different conclusions. If we are not constantly aware of this and make an effort to communicate our assumptions and elicit the assumptions of others, conflict is likely. Each missed communication compounds the problem and escalates the conflict. The illustration of the "Ladder of Inference" gives an example.

We go up the Ladder of Inference when we fail to recognize and communicate our assumptions, inferences, and beliefs at each step, and when we fail to find out what others are thinking. There are three habits we can learn to enable us to climb back down the ladder:

The Ladder of Inference
(read upward)

I had better do it myself	I take **actions** based on my beliefs
You really can't count on anyone	I adopt **beliefs** about the world
I can't trust him with this responsibility	I draw **conclusions**
Bob isn't capable of growing carrots	I make **assumptions** based on the meanings I added
Bob doesn't care	I add **meanings** (cultural and personal)
Bob quit the job	I select **data** from what I observe
Bob didn't finish weeding the carrots	Observable data and experience (as a camera might record them)

1. Make your thinking process visible
What to do
- State your assumptions and describe the data that led to them.
- Explain your assumptions.
- Make your reasoning explicit.

What to say
- "Here is what I think, and here is how I got there."
- "I assumed that . . ."
- "I came to this conclusion because . . ."

2. Publicly test your conclusions and assumptions
What to do
- Encourage others to explore your model, assumptions, or data.
- Refrain from defensiveness when your ideas are questioned.

- Even when advocating, listen, stay open, and encourage others to provide different views.

What to say
- "What do you think about what I just said?"
- "Do you see any flaws in my reasoning?"
- "What can you add?" "Do you see it differently?"

3. Ask others to make their thinking process visible

What to do
- Gently walk others down the ladder of inference and find out what data they are operating from.
- Draw out their reasoning—find out *why* they are saying what they are saying.

What to say
- "What leads you to conclude that? What data do you have for that? What causes you to say that?"
- "How does that relate to your other concerns?"

You can use the Ladder of Inference tool not only with the other primary decision makers on your farm, but also to act more effectively with neighbors and within community, professional, and political-action organizations that are important to reaching your farm's three-part goal.

These aids to developing effective social interaction draw on the work of Peter Senge, Jeff Goebel, and the Taproots seminars. Look in the "Resources" section to find out more about them.

What You Need to Know about
Farm- and Food-System Economics

No decision can truly be economically sound if it is not simultaneously socially sound. Likewise, no decision can be truly socially sound unless it is simultaneously ecologically sound. Thus, unless an economic decision is socially and ecologically sound, it cannot be economically sound in the long run.

—Allan Savory, *Holistic Resource Management*

You do not need to be a professional economist or a certified public accountant to run a farm. But there are a few basic things about farm- and food-system economics that can help make your farm economically viable. Understanding the big picture can also protect you from feeling like a failure when the difficulties you encounter are caused by factors beyond your control.

Cheap Food

Cheap is the basic food policy in the United States. Agribusiness, lending institutions, government, and agriculture schools have banded together to maintain this policy through complex manipulations of price mechanisms, trade rules, and taxes. As a result, middle-income Americans spend a lower percentage of their income on food than citizens of other industrialized countries. The figure for the United States varies from 10 to 15 percent, depending on how it is calculated. Europeans spend closer to 30 percent. The American Farm Bureau boasts about this and sends out press releases in early February celebrating that Americans have already earned enough to cover their food bill for the year. But for the 94 percent of U.S. farmers, organic or conventional, who run the small, independent family farms, this is nothing to celebrate. The constant downward pressure on the pricing of farm products makes it almost impossible for farmers to make a living, and even harder for farm workers.

Food is cheap in this country partly because most of the people who grow, pick, ship, process, sell, cook, and serve the food are poorly paid. According to Eric Schlosser in *Fast Food Nation*, "the 3.5 million fast food workers are by far the largest group of minimum wage earners in the U.S. The only Americans who consistently earn a lower hourly wage are migrant farm workers" (p. 6). The national labor legislation that protects the rights of working people to organize for better working conditions leaves out farm workers. They do not have the legal right to organize or to bargain collectively. By most estimates, 60 percent of the farm workers who harvest the crops on U.S. farms are illegal immigrants, which means they may be subject to mistreatment, underpayment, and other forms of abuse, and they have no legal recourse—except to move on. Tyson is under investigation by the Labor Department for cheating workers out of $340 million in back wages, and for importing illegal workers. If the biggest farms had to pay better wages to farm workers, the price of food would rise and small farmers could earn a living wage.

Food is also cheap because the price we pay for it does not cover many of the costs of producing it, the so-called externalities. When Northeast shoppers purchase Washington State apples, the price does not reflect the expensive system of highways that allows trailer trucks to carry those apples over 3,000 miles. The supermarket price of California vegetables does not cover the cost of the water California farmers use to irrigate them. Even David Pimentel, professor of entomology at Cornell University, with the help of dozens of graduate students, can do no more than *estimate* the costs of pollution of air and water, erosion of soil, and sickness and disease to humans and animals caused by the use of toxic chemicals in agriculture.* Finally, U.S. agriculture's shortsighted dependence on nonrenewable sources of energy does not show up in the price of food.

What Cheap Food Does to Farms

Since the government eliminated parity price supports in 1952, the number of farmers has plummeted from 23 million to 1.9 million, according to the 1997 agricultural census. Parity was a system of adjusting price-support mechanisms to keep farm prices in line with the rest of the economy. In other words, under parity, as tractors got more expensive and workers in tractor factories earned more, the prices farmers received for crops rose so that they could continue to purchase tractors. Under the current system, farm-gate prices do not always even cover the annual costs of production. The 1997 census shows that just under half of farmers claim farming as their chief occupation, and 84 percent say they depend on off-farm income. Black farmers have been squeezed off the land even faster than white farmers: In 1920, 1 out of every 7 farmers was black; in 1982, black farmers counted for only 1 out of 67 and operated only 1 percent of the farms.

Here's an example from the government's own statistics. In 1993, the annual cash expenses for growing an acre of corn in the Northeast (as calculated by the Economic Research Service of USDA) amounted

---

* David Pimental and Anthony Greiner, "Environmental and Socio-Economic Costs of Pesticide Use," in *Techniques for Reducing Pesticide Use* (John Wiley and Sons, 1997); David Pimental et al., "Ecology of Increasing Disease," *Bioscience* 48, no. 10 (October 1998).

to $155.61. The total gross value of selling the corn was $195.69, for a profit of $40.08. However, when the government economists added in the full ownership costs—such items as capital replacement, operating capital, land, and unpaid family labor—the bottom line came out minus $48.95. Oats came out minus $82.83, and milk per hundredweight at minus $2.02.

Despite $26 billion in relief payments, the 2001 aggregate U.S. farm income decreased by 21 percent compared with 2000, and USDA predicted that it would continue dropping through 2002 and 2003. By contrast, agribusiness giant ConAgra's profits rose by 23 percent in 2001, and Cargill Inc. did even better in 2002 with a profit increase of 131 percent.* Of course, 80 percent of the government's agricultural payments go to only 20 percent of the farms.

The Technological Treadmill

Like Earl Butz, secretary of agriculture during the 1980s, the land grant colleges have been telling farmers to "get big or get out." Many farmers have tried to follow this advice, transforming their farms into industrialized agribusinesses. A few have succeeded, and many have failed. Agricultural research for the past fifty years has concentrated on raising yields and equipment productivity per acre, whatever the ecological and quality-of-life costs. Farmers who adopt new industrial technologies early increase their yields and, briefly, profits. As more farmers adopt them, the new technologies bring larger crops—and lower prices to the farmers. Later adopters usually have to go into debt to pay for this "progress," which they believe they have to make simply to keep their gross income from falling. However, when debt interest is subtracted, most farmers find themselves with even less net income, resulting in an ever-descending spiral. Some succumb to the pressures and auction off their farms. Their neighbors expand, buying or renting more land to increase production. The more they produce, the lower the price they receive per unit, so they expand crop acreage to keep the same income: running faster and faster to stay in place. In industry, this is called a speed up.

---

* *AgriBusiness Examiner* and "Corporate Farming Notes," Center for Rural Affairs Newsletter (October 2002), p. 3 (www.cfra.org/newsletter/2002_10.htm).

The structure of the farm economy, with relatively large numbers of farms, but ever fewer buyers of what farms produce and suppliers of what they need, squeezes the farmer from both ends. As the saying goes, "Farmers buy retail but sell wholesale." The biggest farms may be getting bigger, but the farming sector as a whole is losing control to the increasingly consolidated multinational corporations that dominate the food system. Rural sociologists William Heffernan and Mary Hendrickson have been tracking this consolidation. In 1990, the top four beef packers controlled 72 percent of the market. By 2002, that figure had risen to 81 percent. Over that same period, the top four flour millers increased their share from 40 to 61 percent. Only three companies control 81 percent of U.S. corn exports and 65 percent of soybean exports. Cargill is close to the top on all of those lists. Others include Tyson, ConAgra, and Archer Daniels Midland.* When four entities control 40 percent of any market, the Taft-Hartley Anti-Trust Act requires the Justice Department to swing into action. The federal government seems to have forgotten about this law.

The processing and marketing sectors are returning 18 percent on investment, and grabbing ever-larger portions of the consumer food dollar from the farms. Since the beginning of the twentieth century, input suppliers increased their share from 15 to 24 percent, and marketers from 44 to 67 percent. In 1992, Stewart N. Smith calculated that if these trends continued, farmers, as a significant sector of the farm economy, would disappear by 2020.†

The open-border policy promoted by both Republicans and Democrats is contributing to the loss of small farms both in the United States and in other countries. The so-called free trade agreements were written by the ADMs and Cargills of the world, who control much of the trade. Under General Agreement on Tariffs and Trade (GATT) regulations, U.S. agricultural exports have risen 5 percent, but imports have gone up 32 percent. In 2002, this country imported $41 billion worth of food products, and that wasn't all coffee, chocolate, and

---

* "Concentration of Agricultural Markets," paper presented at National Campaign for Sustainable Agriculture Annual Meeting, February 2002, HendricksonM@missouri.edu.
† "Farming Activities and Family Farms: Getting the Concepts Right." Presentation at the Joint Economic Committee Symposium, Washington, DC, 21 October, 1992.

bananas. It included many crops we can produce right here. In 2002, the United States exported $11 billion worth of fruit and vegetables, but imported over $17 billion worth. According to a USDA study released September 3, 1997, the economic impact of the NAFTA on the balance of agricultural trade between the United States and its two neighbors has been a negative $100 million. Mexican tomatoes are underselling Florida tomatoes. U.S. corn growers, on the other hand, favor NAFTA because it opened Mexican markets to them and even helped raise the price of corn 8 cents a bushel. The cost has fallen on Mexico where from 600,000 to a million small corn farmers, who could not compete with the lower price of the imported corn, have been uprooted from the countryside and forced into the army of unemployed in the towns or into the masses of migrants pouring over the border to work on farms in the United States.

If passed, the Free Trade Area of the Americas (FTAA) will drive this trend even further by removing any possibility for localities to favor local producers. The "national treatment" clause prohibits any country from discriminating on behalf of its domestic sector. The "most-favored-nation" clause gives equal access to investors from all FTAA countries. The "performance requirements" section limits a country's right to place performance requirements, such as environmental and labor standards, on foreign investment. And as a final blow against local rule, disputes will be handled by a panel of appointed trade bureaucrats who will be able to override government legislation or force local governments to pay compensation in order to keep local ordinances.*

To keep family-scale, local farms alive, we will have to go beyond planning for individual farms to political organizing on the county, state, and national levels and build alliances with farming groups and movements for social change in other countries too. Food and farming organizations in many parts of the country have begun this work in their local areas and, since 1992, have come together in the network of the National Campaign for Sustainable Agriculture with its focus on federal policy and the Farm

---

* Maude Barlow, "The Free Trade Area of the Americas: the Threat to Social Programs, Environmental Sustainability and Social Justice," A Special Report by the International Forum on Globalization, 1009 General Kennedy Ave. #2, San Francisco, CA 94129 ifg@ifg.org.

Bills.* Vandana Shiva's call for "local food sovereignty" has inspired many efforts around the globe to pass laws that would protect local food production and distribution from foreign competition, remove agriculture from the Free Trade Agreements, and require supermarkets and institutional food services to buy from local farmers.

The Organic Vision

Working for local systems is central to the very definition of organic agriculture adopted in 2000 by the International Federation of Organic Agriculture Movements (IFOAM):

> Organic agriculture includes all agricultural systems that promote the environmentally, socially, and economically sound production of food and fibers. These systems take local soil fertility as a key to successful production. By respecting the natural capacity of plants, animals, and the landscape, it aims to optimize quality in all aspects of agriculture and the environment. Organic agriculture dramatically reduces external limits by refraining from the use of chemo-synthetic fertilizers, pesticides, and pharmaceuticals. Instead, it allows the powerful laws of nature to increase both agricultural yields and disease resistance. Organic agriculture adheres to globally accepted principles, which are implemented within local social-economic, climatic, and cultural settings. *As a logical consequence, IFOAM stresses and supports the development of self-supporting systems on local and regional levels* [emphasis added].†

The vision is out there. Obviously, we have a long way to go to turn it into the universal reality of *community* food security—"a condition in which all community residents obtain a safe, culturally acceptable, nutritionally adequate diet through a sustainable food system that maximizes community self-reliance and social justice."‡

---

* For more detail see "Rebuilding Local Food Systems from the Grassroots Up," in "Hungry for Profit: Agriculture, Food, and Ecology," special issue of *Monthly Review* (July–August, 1998).
† Gunnar Rundgren, "Organic Agriculture and Food Security," IFOAM Dossier #1, 2002.
‡ Definition written by Michael Hamm and Anne Bellows, as cited in Kameshwari Pothukuchi, Hugh Joseph, Andy Fisher, and Hannah Burton, "What's Cooking in Your Food System: A Guide to Community Food Assessment," Community Food Security Coalition, Venice, CA, 2002.

Home Economics–Annual Planning with Long-Term Goals

While it is difficult to have a rapid effect on the macroeconomic forces at work in the world economy, farms that diversify out of commodity production (raw products like milk, corn, or pork typically sold to wholesalers or processors) and sell to local markets stand a chance of doing well.[*] Especially in the Northeast, direct sales from farms to individuals, restaurants, and even schools are flourishing. The number of farmers' markets nationally has risen steadily from 1,755 in 1994 to 3,137 in 2002. USDA statistics show that the 67,000 farms that sell direct are grossing over $1 billion in sales. Since many of these sales are in cash, it seems unlikely that official statistics reflect the real figures. In less than two decades, Community Supported Agriculture (CSA) farms have increased to over 1,000 listings on the Robyn Van En Center Web site www.csacenter.org. Farms have a lot more control closer to home.

When you combine whole-farm planning with available business tools to evaluate your farm's financial status, you can establish the economic base you need to achieve your environmental and quality-of-life goals. Taking the time to keep records and do an annual planning of expenses and income is critical to financial stability. As with goal setting, the more members of your central farm crew that participate in the annual planning, the more creative and realistic your plan is likely to be and the more conscientiously everyone will stick with it.

Record Keeping

To do any kind of planning, you have to start by keeping records. Tossing your receipts in a shoebox is just not good enough. If you have gone through organic certification, you already have many of the records you need: lists of all equipment and what you paid for it, the amendments and pesticides you have purchased, livestock feed, supplements and medications, expenses for seed and production aids such as row covers or plastic. The Internal Revenue Service Schedule F (Form 1040) "Profit and Loss from Farming," provides a useful guide to the categories you can use. Here is the Fed's list of legitimate farm expenses:

---

[*] See Rebecca Bosch, *The Organic Farmer's Guide to Marketing and Community Relations*, NOFA Handbook Series.

- car and truck expenses (form 4562)
- chemicals
- conservation expenses (attach form 8645)
- custom hire (machine work)
- depreciation and section 179 expense deduction not claimed elsewhere
- employee benefit programs other than on line 25
- feed purchased
- fertilizer and lime
- freight and trucking
- gasoline, fuel, and oil
- insurance (other than health)
- interest
    a. mortgage (paid to banks, etc.)
    b. other
- labor hired (less employment credit)
- pension and profit-sharing plans
- rent or lease
    a. vehicles, machinery, and equipment
    b. other (land, animals, etc.)
- repairs and maintenance
- seeds and plants purchased
- storage and warehousing
- supplies purchased
- taxes
- utilities
- veterinary, breeding, and medicine
- other expenses (specify)—Includes investments in land improvement, such as tiling of fields for drainage; gifts and contributions; legal and professional fees; advertising; dues and seminars, including certification; pest control; and others.

Note that the IRS does not include a salary or health insurance expenses for the farmer. In official business tax-think, the farmer receives the difference between gross revenues and expenses, the "profits."

In addition, you need to keep track of the time and money you spend marketing, delivering, and selling your farm products.

Once you are clear on your overall expenses, it's valuable to break them down by crop or separate farm enterprises. To do this, you need to keep ongoing records of how much time and energy you spend on each crop, animal, or project. Some farms maintain a set of notebooks where farmers and their employees daily record the amount of time, seed, other inputs, and then the amount of crop harvested for each field or type of crop (greenhouse, vegetables, field crops, etc). If you produce a value-added product, such as cheese or jam, you need to record the hours spent making it and the money and time spent on ingredients, packaging, labeling marketing, and distribution. Computerized spreadsheets can make this job easier for the computer literate. There are many types of financial software that can work, such as Quicken®, Quick Books™, or simply handwritten double-entry accounting. The Holistic Management Institute (http://holisticmanagement.org) offers free financial software. In many areas, the SCORE program (www.score.org) recruits retired businesspeople who can provide advice to small businesses including farms. At the end of the season, you can review these records to find out which crops are worth growing from an economic standpoint. Farmers like Paul and Sandy Arnold of Argyle, New York, set a goal in dollars per hour: when they find that a crop does not meet that goal, they either find a more efficient approach or stop growing it.

You also need to keep track of all farm income. Organic certification requires that you maintain harvest records and invoices so that you have a clear audit trail for any product you sell. Especially if you are short on money, a bookkeeping system that allows you to track income month by month may be critical to readjusting your course in time to prevent disaster. The conventional way to calculate farm profit is by adding up all costs and subtracting that figure from the sum of all farm income. Costs are divided into fixed costs and variable costs. Fixed costs include things you must pay whether you grow one head of lettuce or ten thousand, such as insurance, taxes, mortgage or rental payments, and so on. Variable costs refer to what it costs you in a given year to grow a given crop: seed; fuel for tillage, cultivation, and harvesting; hours of labor. This conventional system may help you keep in the black, but it does not guide you toward making a profit. As Allan Savory points out in *Holistic Management*, we have a tendency to spend as much as we earn, unless we very deliberately decide to plan for a surplus of income over expenses.

Planning for Surplus

In holistic financial planning, you start by planning your income and then you clamp a ceiling on expenses *by planning your profit first*. The detailed records you are keeping provide you with the data you require to identify where you can make cuts and where you can spend your scarce dollars to greatest effect. Targeting your expenses to fit after the profit is planned requires discipline and the understanding participation of the whole-farm team.

Here is a chart that shows where the income goes in three planning steps:

1. Plan the estimated gross farm income for the coming year.	
2. Plan the profit you need to achieve (or at least to approach) the quality of life you described in your three-part goal.	3. Make the expenses, including both fixed and variable production costs, fit.

As this chart suggests, using your future goals for the farm as your guide, it is important to analyze more than just your fixed and variable costs. You need to separate expenses into costs of production and costs of reproduction. The costs of production include both fixed and variable costs involved in producing crops on an annual basis. Reproduction costs are what it takes to keep the farm going into the future: investments for maintenance, funds for improvements and equipment replacement, training for the farmers, children, and employees, planning for transferring the farm to the next generation, and funds set aside for retirement. If farm revenues only cover the annual costs of production, the farm does not have much of a future.

Allan Savory's holistic management categorizes farm costs in a slightly different way, distinguishing maintenance costs from investments that generate new wealth. Maintenance costs include "all expenses of general administration and other applications that do not directly generate income. Phone bills, office supplies, transport, advertising, supplementary feed, fertilizers, salaries, seasonal help, and your own withdrawings are commonly maintenance costs."

Among the many vocations that human beings follow, farming involves a special trust because farmers work with the forces of nature to create new social wealth. By growing crops and raising livestock well, we increase the wealth in society while keeping a protective eye on the ecosystems, knowing wealth creation has often been a zero-sum, "We-win-you-lose" game with Nature.

Savory divides human wealth—he nicknames it money—into:

- "mineral dollars" (money derived from "human creativity combined with labor and raw resources—soil, timber, dung used as fuel, water, oil, coal, gas, gold, silver, uranium, etc.");
- "paper dollars" (money acquired from "human creativity and labor alone," such as services, financial investing, sports, etc.); and
- "solar dollars," which are produced when we "generate income from human creativity, labor, and constant sources of energy such as geothermal heat, wind, tides, wave action, falling water, and most of all, the sun." Ed Martsolf calls farmers "managers of solar conversion." Creating wealth by the engine of photosynthesis is our very special calling as farmers. By focusing our creativity on how to do this most effectively, we can make our farms economically and ecologically sustainable into the distant future.

So, how do we go about doing financial planning? Top and center, in all your planning and decision making, keep your goals firmly in mind. Then, analyze your farming systems looking for the weak spots. What are the logjams that prevent you from producing more of some crop, or from producing it using less time and energy? Repairing the weak link in your production chain is the first place to spend your time or money. Here are a few examples of weak links. Your crop yields have been low because the plants are not getting enough water. As a short-term solution, you might invest in an irrigation system, while working toward a long-term solution based on increasing the organic matter in your soils through composting and plowing in green manures. If your income from selling crops is limited because there is only one market that controls the price you receive, you might seek a way to diversify your marketing by switching to direct sales, joining a cooperative, or processing some of the crop yourself. Another kind of logjam can be inadequate labor at a critical time in the season, finding yourself stuck doing jobs you do not enjoy, or poor communications among your farm team.

Annual Planning of Expenses and Income

At your annual financial planning session, review your current expenses and sources of income. Look for ways to cut expenses to the bone, especially those that do not generate wealth. If you find that your income is not adequate to meet your goals, you need to improve your sources, or find new ones. A good way to begin such a search is by brainstorming. In a brainstorm, no idea is too absurd. You get relaxed and go wild in listing all kinds of projects—preferably with others. When you have a good list, then you get serious and do a careful analysis of the ideas, culling the ones that do not fit with your goals or do not stand up to the kind of gross-margin analysis described in chapter 6. In making a final choice of any new enterprise, you must be very sure that assembling the resources and doing the work will fit in with your quality-of-life goals.

Elizabeth offers this example from her experience:

At our annual planning this year, my partners and I realized that rising prices of just about everything, but particularly seed (with the new requirement that certified organic farms use certified organic seed where "commercially possible") and fuel, means that we need to raise more income. We also decided to budget more money for labor expenses so that we can be sure to get two interns, pay them better and, if we are fortunate, reduce our work hours slightly.

At least once a week, a well-meaning person suggests to me some additional enterprise we could undertake at our farm to earn more money. We could expand our vegetable production using the many available acres near us, add winter shares, keep bees and sell the honey, grow medicinal herbs, collect wild herbs, raise chickens, hogs, goats, make tinctures, soap, dress up and give tours, and so on. I nod and smile graciously, and think to myself, "and how long would it take me to learn to do that, how much would I have to invest, how many hours would it add to my work year, and do I really need to be any busier?" Greg and Ammie, whose daughter Helen is eight, are even more determined than I am to keep time for their family. So this year, we decided that we would try to increase our income from the work that we are already doing.

Here's how we are trying to do that: Once we had agreed on a budget, we met with our CSA core group to discuss expenses and income with them. (The core must agree to any change in share pricing, number of weeks the program lasts, and number of shares). Before the meeting, we sent them figures showing a range of ways to generate the income we need. We could increase the number of shares, increase the number of weeks the farm supplies shares, increase the share price, or some combination of the three. Greg, Ammie, and I had agreed that we preferred increasing the number of weeks from twenty-six to twenty-seven, raising the price of full shares by $1 a week, but adding only eight more shares. This fits most closely with our focus on improving quality and the feeling of community within our CSA, rather than giving in to the pressures to increase the quantity we produce. At the core meeting, one of our treasurers suggested a raise of 50¢ a week as an alternative, but when we calculated how many more shares we would need to recruit to get the same amount of money, the group came to consensus on the farm's proposal. Although we do not sign a contract, the core members feel committed to recruiting the number of shares we agree upon.

My partners and I are very conservative when it comes to estimating our income for the coming season. Since we grow vegetables, herbs, and flowers, we know that we depend to a great extent on weather conditions. In our four years of production at Peacework, we have had three exceedingly dry years and one cold and wet year. The 2002 season broke records for flooding in June, and there was dry heat from mid-June to mid-September. Clearly, we cannot bet on ideal growing conditions. We have three sources of income: CSA shares, additional bulk sales to members, and sales to stores. We base our budget only on the sale of shares, since that money is guaranteed: our members agree to share the risks with us. If we have a good year, we give them more produce, and if we have a bad year, they still pay us the same amount of money.

In our budget, we include what the IRS considers profit—living wages for our two families, health insurance, and a small pension fund. We also build in an equity payment to me, so that over a period of years, our shares in the farm investment will become equal. The budget covers basic maintenance costs for buildings and equipment, and some money for our continuing education. We do not include capital investments. For purchasing new equipment, we have a special line on the contract with our members for contributions to the farm capital fund. Over five years, members have contributed thousands of dollars to this fund with no strings attached.

Even in a very poor year, we have managed to produce enough food for shares, as well as for bulk sales and some store sales. The money from the latter two sources is over our basic budget. We keep some of this money as a cushion, and spend some of it on capital expenses. While we are not able to plan exactly how much surplus we will generate, our system has reliably generated more money than we planned as expenses. Our members also make part of their payments to us in the fall, to hold their place for the coming season. This serves as our seed money. As a result of our budgeting process, we have no debt.

Sarah Flack, who farms with her husband and parents in Vermont, describes their annual planning process this way:

Since we are two families (me and my husband in one house, and my dad and his wife in another), we each wrote holistic goals in the mid-1990s when we first started studying holistic management. Remi (my husband) and I revise ours annually and have it posted on the wall in our hallway. All four of us got together a couple of years ago and wrote a whole-farm goal by combining the two family goals into one big vision statement for the farm.

Then each winter we do several planning things. First, we draw a visual "map" of all the enterprises and parts of our farm with lines connecting them all together to see how they relate and complement each other ecologically (it's hard to describe; it's a visual thing and fun with lots of colors). Next, we create a biological plan for the next twelve months (modified HM) with lots of fun colors to see how all the animal, crop, and people events on the farm look when spread over the year. This is a great way to see when we need extra help, when we can take a vacation, when we can anticipate losing our minds (June and October— just kidding). Finally, we look at the past year's financial records, start working on taxes, and create a budget for the next twelve months and sometimes look at budgets for the next several years. This I do on QuickBooks with input from the rest of the family.

Monitoring Your Progress

"If you want to make God laugh, tell him your plans," says an old Yiddish proverb. The best laid plans of mice and men do not always produce the results we seek. Events intervene. The economy takes a downturn. A sudden influx of imports eliminates the price premium you expected. A hidden piece of metal pokes a hole in an expensive tractor tire. However, a monthly review of expenses and income will keep your fingers on the pulse of your farm. For many farmers, the arrival of the bank statement each month serves as a reminder to monitor finances. Your planning process will not protect you from change, but it will enable you to keep in touch with what is happening so that you can adjust to realities before things get totally out of hand. A major deviation from your plan may force you to go back to the drawing board and start the process over from the beginning.

five

Choosing Appropriate Tools

In modern times, mechanical and chemical technologies have become the main tools in agriculture. As their long-term damage on farm productivity, environmental health, and even human health and quality of life has become apparent, attempts to make agriculture more sustainable have meant the substitution of more subtle and benign technologies. Another approach that has been taken is to develop and utilize the tools Nature already offers in the agro-ecosystem—that is, to work with natural capital. Farmers are finding this route challenging in the short run because of the ecological knowledge and management skills it requires, but highly rewarding in the long run. Whole-farm planning embraces this perspective, not out of blanket disregard for technology, but to bring the most naturally efficient tools into play.

Nature's tools consist largely of living organisms and their interrelationships. They include:

- *Intercropping.* The practice of intercropping and other sorts of polyculture enriches the immediate environment, affecting fertility, pest management, weather protections, shade, water, etc. One example that deserves more widespread use is the integration of legume crops such as peas or beans for their nitrogen-fixing capacity, with other crops, in alternate rows or within the same row.
- *Bio-fertility agents.* These are microbes and other small organisms that process minerals and other substances into forms that plants, animals, and other microbial life can use. They act as bio-converters living in and on plants, animals, soil, and each other. Using them as tools often means simply encouraging them to thrive in the right environments. One example is mychorrhizae, the fungi that attach themselves to crop roots and extend their reach for nutrients. Mychorrhizae are great scavengers for scarce

phosphorous, which we can encourage by adding compost to our soils. Sometimes one or another of these microscopic bio-fertility agents is missing from your farm's ecosystem and you need to import it to inoculate the soil, or the gut of an animal that lacks it. The California red worm, for example, is a powerful vermi-composter often missing from land that has not seen manure or intensive composting for a while.

- *Biological pest control.* Farmers have begun to use natural enemies to control pests. This particular holistic approach has the potential to become a cornerstone of crop and animal health. You can introduce natural enemies, be they birds, insects, plants, animals, or microbes, but the best long-term strategy is to create an environment that will attract these organisms and cause them to thrive.

- *Grazing.* The use of grazing is one of Nature's less-recognized tools. Using grazing as a tool means using it in a way that builds soil while it grows animal products. The core of the process is a pulsed grazing method that, by properly timing both harvest and recovery periods, keeps pasture in a vegetative state and maximizes its growth over the season. The same paddock is re-grazed several times in a season. After each grazing, portions of the grass roots die back and regrow as the tops recover. The dead root mass becomes soil organic matter. This is a grazing strategy that maximizes both manure and biomass production per acre, creating a surplus over the fertility requirements of the grass. You can fully realize the soil-building potential of grazing by combining it with manure storage and high C/N composting methods that conserve the most nutrients with the least net energy. Pigs, chickens, and red worms have proven to be effective composting facilitators.

- *Beneficial grazing-animal strategies.* Grazing animals can clear brush, remove cover to prepare for seeding, trample in seed to insure germination of overseeded areas, and harvest seed and distribute it to new places through their manure. To use this tool effectively you will often need a high herd density—many animals in a small space for a short time.

- *Integration of crops and livestock.* If there is a single tool that has the most potential to increase efficiency, self-sufficiency, and sustainability of farms, it is the integration of crops and livestock in a single system. There is still a long way to go before agriculture takes full advantage of manure as fertilizer for crops, the non-harvest biomass of the crops as feed for animals, and many other benefits of integrated systems. Alan Nation has said that the best remedy for played-out cropland is a grass rotation, and the best remedy for played out grassland is a crop rotation. Agroecologist Peter Rosset reports that integrated systems demonstrate reduced vulnerability to pests, diseases, and weeds, a lower dependency on external inputs, lower capital requirements, and a greater efficiency of land use.[*] These systems have proven themselves wherever industrial-model specialization of agriculture has not supplanted them, and now scientific research is confirming results.[†]

As we begin to wean ourselves from industrial technology and discover the uses of symbiosis between plants, animals, insects—and especially microscopic organisms—on the farm, it is becoming clear that we are only scratching the surface. The potential is enormous.

---

[*] Peter Rosset, *The Multiple Benefits of Small Farm Agriculture,* Food First Policy Brief No. 4 (Oakland: Institute for Food and Development Policy, 1999).
[†] See, for example, F. Funez, *Sustainable Agriculture and Resistance: Transforming Food Production in Cuba* (Oakland: Food First Books, 2002), chapter 11, "The Integration of Crops and Livestock."

six

A Framework to Make and Test Decisions and Monitor Results

Now that you are equipped with a written description of your whole, your three-part goal, an understanding of the role of the four ecosystem processes in your planning, and an appreciation of the range of tools available, you're ready to put whole-farm planning into practice. Your three-part goal will be your primary navigation instrument in making decisions. A three-part goal develops over time. Using it actively in decision making will help you refine it, make it more specific, and tailor it to better fit the person or people to whom it applies.

Whole-farm planning requires holistic decisions, ones that consider the entire set of conditions and variables that lead to success. Ordinarily, decisions are made considering only part of the full set. For example, when you make decisions only according to their expected effect on the economic "bottom line," you disregard the effects on people's lives, or on the essential environmental functions. Eventually, such actions may backfire. Also, since everything is connected, money-driven decisions, as profitable as they may seem in the short run, will feed back through the system, affecting personal quality of life, environment, neighborhood relations, market economy, and so on. Thus actions that initially boost profits may hurt them in the long run.

For example, suppose you decide to increase lettuce production. The first season you sell more lettuce and make more profit. But because of the strain on the existing labor, the best workers quit. The next year your labor is less reliable and your profits go down. Or since lettuce is a heavy feeder and you haven't adjusted your fertility program, several years later the farm's overall fertility is down and so are profits. Or, profits are up, but not as much as they might be if you had spent the time on better lettuce

In complex systems, cause and effect are distant in space and time.

marketing instead of increased lettuce production. Or there is another enterprise on the farm that is more efficient than lettuce at making profit, so it would have paid you more to put your efforts there. Or, you hadn't stopped to think that you don't enjoy growing lettuce much anyway, and prefer your salads made with kale, so eventually your management starts to slip and so do profits!

To help you consider the range of variables involved, in this chapter you will learn how to subject your decisions to a number of testing questions. You should ask these questions after you have consulted all the usual sources: your own experience, other farm decision makers, experts, farm records, books, and so on. In other words, they are a final check. Depending on the type of decision, some of the tests may be relevant and useful, others not.

- **Does this decision move the farm toward or away from my goal?** You should ask this question of every decision. Ask it first and again at the end of your testing as a test of your gut feelings on the decision. Ask it in three ways:
  1. *Values.* Does it help bring about the quality of life that you value?
  2. *Forms of production.* Does it help you succeed in the things you want to do on the farm and in life?
  3. *Sustainability.* Will it lead toward or away from the future resource base you have described? Some things in this part of your goal may seem unattainable in any reason-

able time, but many small actions in the direction you want to go can add up. Remember that your future resource base includes both environmental considerations and people: Will this action support the kind of neighbors and improve the rural communities and community services you desire?

- **Does this action get to the root cause?** If the decision involves an action to deal with a *problem*, you should ask if the decision addresses its root cause, or just a symptom. If the problem keeps coming back, you have probably addressed a symptom, not a cause. It sometimes helps to ask: "How could my action be refined to even more effectively deal with the root cause?" To help you find the root cause, you can ask "the Five Whys." This is just an easy way to remember to keep asking *why* until you get to the bottom of the question. For example, suppose your lettuce crop is going to bolt and the decision is to cover it with shade cloth. Here is a hypothetical application of the Five Whys.
  Q: Why is it bolting? Is it really the lack of shade?
  A: No, it's the wrong variety for that time of year.
  Q: Why did that variety get planted then?
  A: The person responsible for planting did not inventory the seeds on hand in time to reorder the necessary varieties.
  Q: Why is that?
  A: The planter had too many other responsibilities around reordering time.
  Q: And why is *that*?
  A: The committee that planned the seasonal division of labor did a poor job.
  Q: And why is *that*?

So far "the Five Whys" process has revealed that the lettuce-bolting problem is perhaps only a symptom of a social organization problem that possibly requires a reorganization of work responsibility on the farm. You may still decide to use shade cloth on the lettuce as a quick fix, but the seductive quick fixes tend to be technological, and usually treat only symptoms. The root-cause test directs decision making to deal with a deeper cause, and addressing that is likely to yield better, more permanent results.

- **Does this action address the weak link?** Your farm consists most importantly of chains of relationships that need to be strong and working smoothly. At any given time you can find one place in a chain that is the weakest, so that an improvement you made there would give you the most reward for your effort. There are three chains that are most essential to good management. The *social chain* links people who are important to the success of the farm. Whenever your decision involves this chain you should test whether and how it affects the weak link. The *biological chain* links organisms, parts of organisms, and stages in the lives of organisms. Look for weak links here when your action concerns the health of organisms you desire on the farm, or control over organisms you don't want. The *financial chain* links all the stages of production, from your initial investment in raw materials to marketing the product. It includes creation of the raw product, and processing, packaging, transport, and selling activities. You should test any action that involves financial profit for how effectively it attends to the weak link in this chain. Finding that link helps you get the most bang for your buck.

To see if the action you plan addresses the weakest links in each chain, ask:

1. *Social*: Does this action deal with any confusion, anger, or opposition it could create in people whose support is needed in the near or distant future? How could you enhance this action to further strengthen relationships, collaboration, enjoyment, etc.?
2. *Biological*: Does this action address the most vulnerable point in the life cycle of a pest organism or the weakest point in its system? How can it most effectively enhance the organism you are seeking to protect?

3. *Financial*: Does this action strengthen the weakest link in the chain of production?

Let's take an example through these weak-link tests. Suppose you want to spray an apple orchard that is yielding many wormy apples. This action has potential impact on all three chains. To test the social chain you ask: How will the person who has to do the spraying feel? How will the neighbor who receives the spray drift feel? How will the customer feel? If your answer to such questions reveals weak links in the social chain, you may decide it fails this test. Then you ask: Does spraying attack the weakest point in the life cycle of the pest? Let's say you decide that the weakest point is the mating process. Then you might consider using pheromone traps instead to disrupt mating behavior. Then you ask: Is worminess in the apple production the weakest point in the conversion of apple orchard production to dollars? If your orchard needs a lot more investment to get a clean crop, you might decide that there is more profit right now in not spraying, and making the apples as they are into cider. So your weak link is not in apple production, but in processing. In this example, the decision to spray fails the test in all three areas.

- **What's the best source and use of money and energy for this action?** On the subject of money, most farmers would agree that it is better to stay out of debt. You may even have included freedom from debt in your three-part goal. Yet many farmers carry substantial debt loads that reduce profits. Money from production credits and subsidies becomes a habit, and when no longer available the withdrawal effect can be devastating. When asking the important question, "Will the money for this action come from your earnings or from a loan," it helps at the same time to ask: Will the money be used for running expenses, and be merely consumptive with no lasting positive effect? Or will it be an investment that actually generates income? Sometimes a bank loan can be justified if it will be used in a way that improves farm income in the future.

In terms of energy, we now know that fossil fuel use has many downsides, especially for future generations. What does your goal say about the best source of energy to use on your farm? Does the action you propose agree in terms of energy source? If your goal is sustainability in regard to energy, what sources of solar energy could be used for this action? Or developed for eventual use? As with the money test, ask at the same time: Will this action use energy for lasting improvement?

According to Dick Levins of the Land Stewardship Project, in a recent year, farmers generously rewarded nonfarm, mostly nonlocal corporations with $50 billion to provide machinery and maintenance, petroleum-based inputs, and interest on loans to pay for them. He says the level of these expenses on your farm "is one indicator of how willing a farm is to share its income with non-farm corporations. Expenses accounted for by chemicals, commercial fertilizer, and gas-guzzling equipment are also a measure of how a farm is interacting with its environment."*

- **Which action, in terms of your three-part goal, brings the greatest return for the time and money spent?** Ask this question when you are considering two or more actions and want to compare them. It is like the weak-link question in that it asks which of the actions under consideration best attends to what is currently the weakest part of an enterprise. It is a question that asks which action will give the most marginal return, the most bang for the buck. But it is about more than economic return. This testing question is quite broad and includes considerations that are not tangible or quantifiable. The following example shows how non-economic concerns come into play on Peacework Farm:

For years, we have been doing all of our transplanting by hand. After suffering from sore knees for an entire season, Ammie proposed that we purchase a waterwheel transplanter. Is buying a transplanter a decision that fits with our farm goals?

Let's see: We already have a greenhouse and produce our own transplants, and we have the use of a tractor with creeper gear

---

* *Monitoring Sustainable Agriculture with Conventional Financial Data* (Land Stewardship Project, White Bear Lake, MN, June 1996, http://www.landstewardshipproject.org/mtb/monitoring.pdf).

and enough horse power to run a transplanter. We have enough cash in our capital fund to pay for one without encumbering ourselves with debt. All of our knees are getting older. If the equipment runs well, use of this machine will shorten our work hours. On uneven ground like ours, a waterwheel transplanter allows the operators to firm the plants into the soil by hand. In terms of water needs, the transplanter is an improvement over hand transplanting since it starts the plants out with enough water that you do not have to start irrigating immediately if conditions are dry. However, the transplanter transforms a quiet job into a noisy one requiring more passes over the field with a fuel-guzzling tractor. On balance, after observing the equipment of several farming friends, we decided that the time and body savings outweigh the negative environmental costs. In the long run, we hope to convert the diesel tractor to biodiesel, substituting waste vegetable oil for petroleum, so the impact on air quality will be reduced.

- **Which enterprise contributes most to covering the overhead of the farm?** This question is most important to annual planning, to assist decision making about which enterprises to change, continue, or eliminate. But it may also be relevant if you are considering changes to enterprises during the year. To help you decide which should get more of your effort, you can do a gross-profit analysis. The annual gross profit for each enterprise is its gross income less only the expenses that were specific to that enterprise. In other words, it does not include expenses paid just to keep the whole farm running, or overhead. Gross-profit analysis compares enterprises in terms of the income they produce per unit of resources that are used by all enterprises, but may be in particularly short supply in a given operation. Examples of resources that may be scarce are acres, dollars for capital investment, labor hours, etc. For example, a gross-profit analysis for labor (such as that shown in table 1) tells you what income is expected per hour of labor for each enterprise.

Table 1. Example of Gross-Profit Analysis						
Gross Profit Analysis	Skins	Handknits	Yarn	Meat	Cheese	TOTAL
Gross Income	2,110	427	3,457	2,734	15,696	24,400
Expenses above Overhead	1,184	–	1,594	1,043	1,255	5,076
Gross Profit	926	427	1,863	1,691	14,441	19,348
Hours of Labor	47	50	48	43	1,863	2,051
Return/Hr. of Labor	19.7	8.54	38.81	45.6	7.75	

As detailed in table 1, suppose your farm has five enterprises, producing sheepskins, handknits, yarn, meat, and cheese. The farmers are working themselves to the bone making all these products. You want to know where their labor is yielding the most profit per hour. Since the analysis shows cheese to yield the least dollars per hour of labor in the previous year, you may want to consider putting that labor into an enterprise other than cheese, in order to get more profit from your scarce supply of labor. But a decision to eliminate or cut back the cheese-making enterprise needs to take into account all the values in your three-part goal, not just financial expectations. And it needs to consider the indirect impacts that might have on the profit of other enterprises over the long term. In this case, it so happens that (*a*) the decision makers really like making cheese, as stated in their holistic goal; (*b*) the cheese acts like a loss leader in supermarkets: it is what makes the farm distinctive and visible in its market, and therefore boosts sales of all the other products; and (*c*) currently cheese income represents over half of total farm income. Therefore the farmers might well decide to keep making cheese even though its labor-intensive nature is less profitable per hour, and eventually double production in order to gain labor efficiencies of scale. But the gross-profit analysis gave them a better picture of farm profits to use in their decision making.

These tests may take time at first. But with practice you should be able to run through them quickly as a final holistic check after all your other thinking about a decision is done. On the following page is a worksheet you can copy and use for any decision. For each question there is a place to mark whether the decision passes, fails, is questionable, or is not applicable.

DECISION-TESTING WORKSHEET				
Question	Yes	No	?	N/A
Does this decision move the farm toward or away from my goal?				
Values. Does it help bring about the quality of life that I value?				
Forms of Production. Does it help me succeed in what I want to do on the farm and in life?				
Sustainability. Will it lead toward or away from the Future Resource Base I have described?				
Does this action get to the root cause? Or just deal with a symptom?				
Does this action address the weak link?				
Social: Does this action deal with any confusion, anger, or opposition it could create in people whose support is needed in the near or distant future? How could I enhance this action to further strengthen relationships, collaboration, enjoyment, etc.?				
Biological: Does this action address the most vulnerable point in the life cycle of a pest organism or the weakest point in its system? How can it most effectively enhance the organism you are seeking to protect?				
Financial: Does this action strengthen the weakest link in the chain of production?				
What is the best source and use of money and energy for this action?				
Money				
Source				
Use				
Energy				
Source				
Use				
Which action, in terms of my three-part goal, brings the greatest return for the time and money spent?				
Which enterprise contributes most to covering the overhead of the farm?				

Remember, if your decision deals with a problem, ask the root-cause question first. It is often useful to run sheets to test and compare several possible actions at the same time.

As we explained in our introduction, complex systems like farms are in constant change. This means that a decision that was right at the time you made it may not work later. Or you discover it was wrong in the first place. If you took action to address a problem, monitoring the result can suggest whether you addressed the root cause of the problem. Some changes you monitor are easy to see, like weather, fast growth in plants, or the movement of birds, insects, and animals. Some you will need to observe over a season or several years in order to discern changes. Soil metabolism slowly consumes organic matter, for example. Slow changes in soil composition, ground cover, and populations of pests, beneficials, and pasture species sometimes require special monitoring tools. By contrast, the genetics of some of the smaller species of organisms can change rapidly enough to surprise farmers. Insects that mutate to become immune to Bt are one example. Then you may want to change the genetics of the plants and animals you grow to make them more sustainable under your farming conditions. You will need a genetic selection and monitoring process for that.

The nature of the whole you are managing is such that the sooner you can recognize deviations from the results you want, the less it will cost to bring things back under control, or even replan if necessary. Monitoring financial matters such as monthly cash flow was covered in chapter 4. The Land Stewardship Project offers a "Monitoring Toolbox" that is especially good for tracking all sorts of biological changes on your farm. Check the "Resources" section at the back of this manual for this and other organizations.

Testing and monitoring your decisions this carefully may be new to you. In order for it to become a habit that you do automatically in your farming life, it helps to form a learning community with other farmers who are learning the same tools of whole-farm planning. When the minds of the whole group test the decision, hard work is turned into fun. A group will bring a wider range of thinking to bear, and often come up with alternatives you had not thought of. An experienced whole-farm planner may facilitate the first sessions of a learning community, but essentially it is farmers helping each other learn.

seven

Conclusion:
Keeping Our Balance

We eco-farmers set the standard for balance. We decry the no-impact worldview of Wall Street. We must make consistency and balance normative in both words and action. Our world needs us to provide examples of balance, to show that production need not compromise the local ecology, to show that a profitable business need not adulterate the demographics of the community. We need to pioneer new ways of growing that regenerate communities and our families rather than destroy the bedrock institutions of a culture. If we don't, who will?
—Joel Salatin, "Balance: Stability for Your Life and Farm"

An article of faith for businesses in these United States asserts: "If you aren't growing, you're dying." The agricultural establishment bullies farms with the demand: "Get bigger or get out." There is so much pressure in our culture to think this way that it takes a major effort to resist. We hope that this manual will serve as a guide to an alternative way of thinking and decision making. For our farms to be sustainable, farmers must be free to consider all possibilities. There may be times when expansion is appropriate, but to balance our personal, environmental, and economic goals, we may decide that it makes more sense to keep the farm the same size or even make it smaller. Our mantras should be "dynamic equilibrium" and "resilience."

Achieving dynamic equilibrium means finding a way to run your farm so that it does not run you into the ground. Designing your farm for maximum resilience means that you can withstand the shocks of bad weather, adverse business cycles, and the fragility of human existence. Most of us who are doing market farming chose this path because growing vegetables or flowers or tending livestock is something we love to do. We get satisfaction from living close to the earth, working outdoors, planting and seeing things grow, nurturing living creatures, using our bodies as well as our minds. We can imagine small-scale farming as a wonderful style of life for ourselves and future generations. The trick is to design our farms so that we do not destroy our love. That means we have to find the right scale of activities, the number of acres we can handle, the optimum amount of equipment, the fairest markets, and the financial goals that will make our farming socially as well as environmentally sustainable.

Students of resilience in ecological systems have identified three basic elements that can help us define resilience for a farm: the capacities to absorb change, to self-organize, and to adapt. Let's look at what this means:

Absorbing change. There are many kinds of buffers that enable a farm to withstand changes, whether gradual or sudden. Here are a few examples. Managing soils for higher organic matter and soil health is a buffer against uneven rainfall. If there is a heavy downpour, soils with decent levels of organic matter absorb and retain the moisture so there is little runoff and erosion. In a drought, the spongelike quality of the soil continues to supply water to crops. Farm design can help a farm deal with the behavior of neighbors. A certified organic farm maintains physical barriers like hedgerows, tree lines, and uncropped buffer zones around its fields so that when a nearby conventional farmer sprays a pesticide, the damage is limited. Building a farm team that has redundancy of skills reduces dependence on any one member, so if someone is injured, gets sick, dies, or leaves, the others can carry on whatever jobs need to be done. A diversity of crops cushions a farm against the vagaries of the weather and also markets, so that a farmer is not forced to be a price taker, dependent on only one or two buyers.

Self-organization. In following a farming plan based on wholes and goals, there are many ways in which we are likely to be organizing networks of support that will cushion us from outside pressures. By cooperating with other people, we gain skills in communication, build trust, and learn how to decentralize power from the big institutions of our society that have become increasingly difficult to control. Through direct sales at farmers' markets, farm stands, or community supported agriculture, farms develop long-term relationships with loyal customers who vote for the farm with their food dollars. Some CSAs have core groups of members who make a much greater commitment to the farm, helping recruit members, harvesting and distributing the crops, giving moral support through hard times, and even lending or donating money. By forming cooperatives with other farms, we achieve greater strength in the marketplace and better prices than selling or buying on our own. Farming organizations like NOFA serve us on many levels as social and informational networks, enabling us to teach one another, farmer to farmer, without waiting on outsiders to provide what we need. As farmers, we also help organize the living communities of microorganisms, plants, and animals on our farms so that we become more self-sufficient in fertility, feed, and energy. By understanding nutrient cycles, growing cover crops and forages, making compost, harnessing wind, water, and solar power, we reduce our dependence on purchased inputs and fluctuating prices.

Adaptive capacity. By closely observing our farms and learning to read the many signals of the changes underway, we can respond quickly and effectively. The underlying philosophy of organic and biodynamic farming guides us to think in terms of cycles as well as to an awareness of both the social and ecological impacts of our farming. As organic farmers, we are managers but also students of the natural systems of which we are a part. By facing the reality that we never have all the information we need and that our decisions may be wrong, we can learn to be alert to the earliest signs of feedback telling us we need to adjust our course.

Managing for sustainability is a balancing act. In farming, we are afloat in a sea of uncertainties about weather, markets, pricing, and our human community. Unpredictable events break over us from the outside world. Our own objectives and desires shift and swirl. In the face of all this, we must somehow be persistent without becoming rigid. Like a champion boxer, we have to roll with the punches. A business that is constantly expanding, feeding like a voracious predator on the failures of its neighbors, resembles a cancerous virus. A resilient farm that is in dynamic equilibrium is more like a brilliant, colorful whirling top. The holistic planning and decision-making process we recommend in this manual is based on sensitivity to and monitoring of the changes within us and around us. It offers us a way to systematize our instinctive sense of the ways of our universe so that we can farm for a healthier present and a more peaceful future.

eight

Examples of Whole and Goal Statements from Practicing Farmers

Katie Smith: Two Examples

After reading *Working Solo* by Terry Lonier, I followed chapter 3's "Charting a Business Plan" to develop my own for Kline Kill Organic Gardens (KKOG).

Kline Kill Organic Gardens Business Plan, 1995

Purpose

The purpose of this business is to produce healthy organic food for local and regional people. This business will provide me a full-time living without having to obtain work in the off-season.

Business

KKOG is a 2½ acre market garden located in Chatham, New York, on land rented from The Berry Farm. It produces certified organic vegetables to regional and local outlets. At this time the farmer seeks to spend as much time as possible on the farm to maintain control over production. The hope is that after establishing markets and consistent production Kline Kill Organic Gardens will become a CSA farm.

Marketing

KKOG will sell to The Berry Farm, six local restaurants, and the Omega Institute.

Production

The production process will begin with the purchase of a greenhouse, irrigation equipment, Buddingh cultivator, and delivery vehicle. I will rent The Berry Farm's tractor, irrigation pipe, mower, and rototiller at a fair rate with the long-term vision of buying my own tractor. I will work with one part-time person to produce, harvest, and deliver vegetables.

Chart Your Timetable

Timetable based on growing season.

Finances

First Pioneer farm credit for $10,000 capital loan and $5,000 operating loan

Capital Expenses

Van	$5,000	
Greenhouse	$2,000	
Irrigation pump and drip equipment	$800	
Fence & charger	$600	
Basketweeder	$600	
Total		$9,000

Operation Expenses

Equipment rent	$200	
Tractor rent	$400	
Land rent	$160	
Seeds	$500	
Supplies	$1,500	
Workers comp (did not do)	$1,200	
Employee	$3,000	
Fuel	$500	
Misc. (registration, bank fees, etc.)	$800	
Total		$8,260

Personal Expenses

Rent & food	$3,600	
Insurance	$1,500	
Misc.	$1,500	
	$6,600	
Total expenses		$23,860

Income Projections

I had 5 main crops and researched markets and prices.

Lettuce	$7,000
Greens	$6,000
Tomatoes	$3,000
Leeks	$3,000
Other	$5,000
Total	$24,000

Reflections Nine Years Later

The process is more important than the accuracy of the plan. Thinking through all of the possibilities and researching costs and markets gave me a solid understanding of the business that I was getting into and what I would need to do to make it a success.

The Farm at Miller's Crossing (Transition Year 2000)

This plan follows the principles of holistic resource management discussed in chapter 4.

Quality-of-Life Statement

Express true motivations: convictions, beliefs, values, and moral characteristics— What do you want to be?

We would like to raise a family here on this farm and have time to enjoy each other year round. We would like to create a cozy and private home and a beautiful farm. We would like to travel if we desire to during the winter. We will be good stewards of this place. The barns and build-ings on the farm will be improved and their productivity increased. All of the land will be managed well and appropriately and will stay in agricul-tural production. We will contribute to and be a part of a thriving local economy here in Columbia County.

Forms of Production

Identify the activities that you need and wish to use to produce and sustain that quality of life—What do I want to do?

Profit from market garden. Recreational time and fun scheduled during the season. Create recreational space in barn for friends and family and public space for CSA members and other people interested in the farm. Rent land for field crops or collaborate with local growers on beef, pork, or sheep production. Profit from those ventures. Buy local products whenever possible and work with organizations promoting Columbia County Products.

Future Landscape

Describe your vision, your projection of what will be needed (landscape, community, family, business, etc.) to support the quality of life indefinitely—What do I want to accomplish?

Better knowledge, and therefore use, of cover crops to improve fertility and microbial life in our soils and to increase disease resistance in our plants; use beneficial insects and animals to decrease pest problems and slow the spread of disease; an increase in all of our produce markets; animals grazing the grass that's growing, and markets to sell those animals; a full-time employee who is committed to the farm and shares our vision. Buildings fixed, maintained, and made more productive. Public bathroom put in as well as apprentice housing somewhere on farm.

Finances

Capital loan $30,000; operating loan $6,500

Capital and Operating Expenses

Chisel plow	$1,075
Cooler	$4,521
Disk	$800
Fence & charger	$231.71
Germination chamber	$1,400
Irrigation pump	$620
Kubota 6800 (trade-in)	$17,100
Mower	$1,853
Perfecta	$1,950
Scale	$465
Speedy seeder	$1,332
Transplanter	$1,819
Water pump	$480

Water tank	$344.56
Greenhouse ('99)	$10,000
Capital Expenses	$43,991.27
Operating Expenses	$72,857.75
Personal Salary	$26,000
Total Expenses	$142,849.02
Projected Income	$105,000

Kaylie and Abe's Holistic Goal

This plan followed the guidelines found in Whole Under Management

Decision Makers
Kaylie and Abe

Resource Base
We have our knowledge and experience, reputations, and drive to build a good life together.

People
Our immediate and extended families
Our farming friends
All of our other friends
Neighbors
Organizations, including NOFA, Vermont Grass Farmers, USDA, NRCS, UVM Extension, FSA, ATTRA, White River Partnership, The Savory Center, HM Educator Network, etc.
Agricultural consultants
Feed, livestock, and equipment dealers
Our customer base, including carpentry, knitting, and meat customers
The communities of the White River Valley and Williamstown area

Land and Home Base
Our house
The solar barn

Sheds, trailers, etc.
Our area of the farm
Land-use agreements with neighbors

Equipment and Livestock

Computers	Furniture and kitchen
Abe's tools	2 guitars
2 knitting machines	Our books
Sewing machine	Lots of barrels
Weaving loom	Flock of sheep
Sheep equipment	1 goat
2 freezers and a fridge	2 trucks
2 cats	1 car
Chicken plucker	Field mower, trimmer
Chicken tractors	1 sheepdog

Access to:

Dad's tractor, tools
Mike's equipment
Dad's and Mike's shops
Church kitchen

Money:

Income from carpentry
Income from Kaylie's misc. jobs
Labor for goods/service exchanges

Savings

Income from sheep and poultry
Cost share programs, FSA, EQUIP, etc.
Possible loans from family
Loans from bank

Holistic Goal

Quality of Life:

We are madly in love with each other and value:

- Romance
- Being with each other in a respectful, supportive, and joyous way
- Feeling secure in our relationship
- A stable home environment
- Open, trusting, and honest communication
- Good health
- Growing, cooking, and eating healthy food together
- Farming
- Study, learning, and teaching
- Creative time beyond livelihood
- Farming and working with our hands and minds
- Being usually at the homestead and in the countryside
- Being well compensated for our labor
- Being financially secure
- Family and new and old friends
- Connections with other farmers
- Community service
- Spending time with youth
- Social justice, healthy land, healthy community
- Strong local economy

Forms of Production:

- A loving, growing, joyous relationship
- We need to manage our time to allow for rest, study, and play
- One day of rest a week
- Accounting system, daily and weekly routines and habits
- An annual financial plan
- Grazing and land plan
- Products that meet peoples' real needs
- A local customer base and good relationships with clients, neighbors, and service providers
- Healthy, productive, and beautiful landscape with increasingly fertile soils
- Most of our own food
- Participation in farmer networks
- Contribution to community

Future Resource Base

- We need to be seen as a happy couple, respectful, reliable, and honest; meeting peoples' needs; community oriented; and as an example of ecological, profitable farming.
- A robust land base with a foundation of perennial polyculture with pasture, forest, and clean water bodies.
- A strong water cycle with water available for soil life and wildlife, drinking, livestock, irrigation, and recreation.
- A strong, rapid mineral cycle with fully covered, high-humus soils and balanced minerals.
- High biodiversity, with lots of edge and good habitat, a diversity of domestic and wild species.
- High energy flow through pasture, cropland, forest, and constructed microclimates (greenhouses) that will nourish us, soil life, and wildlife, and can be harvested as solar dollars.

Our community will have a strong local economy, be stable and thriving, with educated, happy people who value good local food and neighborliness. The community will have a broad range of local services, including a great school, good library, active farmers' market, needlework shop, banking, vehicle repair, etc.

We will need access to good slaughtering, fiber-processing, and marketing facilities.

All of our power and fuel needs will be met with renewable sources. Products and services we use will be produced ecologically and respectfully of the rights and needs of producers.

Richard Wiswall

[My] two goals are not that different. One is from 1996 and goes: I want a healthy life with family and friends, earn a living that I have passion for and that doesn't compromise my values of a healthy planet, community, and family. I want to be able to afford to travel, not work very long hours, have money enough not to worry about it, feel more spirituality, play more music, spend more time with family and friends.

The second one from 2001 goes: I want to feel free to do what I want without worrying about money, enjoy the outdoors (hiking, biking, canoeing) read books, help others, maintain friendships, be able to relax without feeling antsy, spend lots of quality time with the kids, and help them along, teaching them whenever appropriate. Have fun with the kids. Travel and explore the world. Maintain and increase my health.

Resources

Holistic Approaches to Farming

Here are resources that take an explicitly holistic or comprehensive approach to whole-farm planning.

Organizations
Allan Savory Center for Holistic Management
1010 Tijeras, NW
Albuquerque, NM 87102
Phone: (505) 842-5252
Fax: (505) 843-7900
E-mail: chrm@holisticmanagement.org
www.holisticmanagement.org.

  Holistic Management, Second Edition, by Allan Savory with Jody Butterfield (Washington, DC: Island Press, 1999) is the current bible on holistic planning and decision making. Even though the book focuses on problems of farming in the arid Southwest, we have drawn heavily on this work to adapt its insights to the Northeast environment.

  Financial planning software is available from the Center for Holistic Management, and there is a quarterly magazine, *In Practice*, that shares experiences of people all over the world who are following the principles of holistic management.

The Minnesota Project
1885 University Avenue West, Suite #315
St. Paul, MN 55104
Phone: (651) 645-6159
www.mnproject.org/

  The Minnesota Project is a network connecting many whole-farm planning efforts in the Great Lakes region, and an avenue to many other resources on the subject. Their newsletter, *The Whole-Farm Planner*, describes farmer experiences with whole-farm planning. There are many case studies in the newsletter archives.

Land Stewardship Project
Twin Cities Office
2200 4th Street
White Bear Lake, MN 55110
Phone (651) 653-0618
Fax: (651) 653-0589
www.landstewardshipproject.org

The Land Stewardship Project publishes *The Monitoring Toolbox* and *Monitoring Sustainable Agriculture with Conventional Financial Data.*

Great Lakes Basin Whole-Farm Planning Network
www.misa.umn.edu

This network offers many resources, including a list of aspects that a whole-farm plan should address, including

> farm goals
> economics
> water quality and management
> soil conservation and management
> pest management
> crop rotations
> tillage
> health and safety
> grazing
> woodlot management
> energy efficiency
> noise and odor
> fish and wildlife management

ManagingWholes.com
www.managingwholes.com/good-goals.htm

This Web site contains lots of jargon-free language about goal setting and other practices and tools of holistic management on farms.

The Taproots Seminars

These seminars are week-long workshops on how to build effective organizations. They are an excellent way to learn the relevant social skills and processes. They are a Learning Communities Project of the Center for Sustainable Systems, P.O. Box 342, Hartland, VT   05048. Phone: (802) 436-2333.

Pasture Management

pasturemanagement.com

This excellent Web site is run by Wayne Burleson. A source of good stories and photos about people solving conflicts and managing wholes in Montana and elsewhere. Burleson has also written a workbook on holistic peacemaking and problem solving.

Whole-Farm Planning with Holistic Management

www.umass.edu/umext/jgerber/hmpage/hmpage2/mainpage6.htm

John Gerber has produced an interactive Web site to help people quickly start the process of creating their own description of their farm's Whole and Three-Part Goal. He has graciously allowed us to use some of the materials on his site in writing this manual.

Goebel and Associates

www.aboutlistening.com

Jeff Goebel's Web site has good material on consensus building and community development and stories about particular cases of both.

Books

Funez, Fernando, et al. *Sustainable Agriculture and Resistance: Transforming Food Production in Cuba,* (Oakland: Food First Books, 2002). The importance of this first major report on the Cuban experience is that it describes sustainable practices that go far beyond what are common in the United States.

Jacobs, Jane. *The Nature of Economies,* (New York: Vintage Books, 2001). Since human beings "exist wholly within nature as part of natural order in every respect," Jacobs draws upon natural processes for clues to understanding human economies. Her book is in the form of a dialogue with examples that are easy to understand.

Rosset, Peter. *The Multiple Benefits of Small Farm Agriculture*, Food First Policy Brief No. 4 (Oakland: Institute for Food and Development Policy, 1999); www.foodfirst.org. Peter Rosset is one of the few thinkers who understand the potential of holistic, integrated farming systems.

Senge, Peter M., *The Fifth Discipline: The Art and Practice of the Learning Organization* (New York: Doubleday, 1990), and *The Fifth Discipline Fieldbook: Strategies and Tools for Building a Learning Organization* (New York: Currency, 1994). These Senge books are a major resource to learn how to collaborate effectively with other people. Our presentation of the "ladder of inference" tool to prevent and resolve conflict is a strategy is drawn from the Senge *Fieldbook* via Carol Anderson, who teaches in the Taproots Seminars.

Alternatives to Mainstream Business Models

Organizations
The E. F. Schumacher Society
Box 76, RD 3
Great Barrington, MA 01230
Phone: (413) 528-1737
www.smallisbeautiful.org

Susan Witt, continuing the work initiated by her deceased partner Robert Swann, provides advice and information on the many ways that small can be beautiful: local economies, land trusts, local currencies, sample leases. The society has an excellent library that is open for public use.

A Whole New Approach
Ed Martsolf
1039 Winrock Drive
Morrilona, AR 72110
Phone/Fax: (501) 727-5659
E-mail: ed.martsolf@mev.net

This organization provides training in whole-farm planning.

Books

Berkes, F., and C. Folke, eds. *Linking Social and Ecological Systems: Management Practices and Social Mechanisms for Building Resilience* (Cambridge: Cambridge University Press, 1998).

Berkes, F., J. Colding, and C. Folke, eds. *Navigating Social-Ecological Systems: Building Resilience for Complexity and Change* (Cambridge: Cambridge University Press, 2002).

Daly, Herman E. "Sustainable Growth? No Thank You." In *The Case Against the Global Economy*, ed. Jerry Mander and Edward Goldsmith (San Francisco: Sierra Books, 1996).

New Society Publishers (4527 Springfield Ave., Philadelphia, PA 19143) offers resource books on group process and decision making, such as *Manual for Group Facilitators, Resource Manual for a Living Revolution*, and *Democracy in Small Groups*.

The writings of E. F. Schumacher, such as *Small Is Beautiful: Economics As If People Mattered* (Hartley & Marks: Seattle, 1999), as well as those of Richard Borsodi, George Benello, Hazel Henderson, and Barbara Brandt delve into the theory of small-scale economics, local economies, and cooperative enterprises.

Other Farm-Planning Resources

There are many other resources that focus on one or another aspect of farm planning, but not always on the whole. Here are a few we have studied that were designed primarily to address environmental concerns.

Ontario Environmental Farm Plan

Ontario Federation of Agriculture
491 Eglinton Avenue, W., Suite 500
Toronto, Ontario M5N 3A2
Contact: Ms. Lynn McNiven

Ontario Soil and Crop Improvement Association

Box 1030, 52 Royal Road (OMAFRA Bldg)
Guelph, Ontario N1H 6N1
Phone: (519) 767-4608

Ontario Environmental Plan is a detailed scoring system for evaluating environmental strengths and weaknesses of a farm. The workbook, with twenty-three worksheets, takes you through your farm, starting at the farmstead and then proceeding field by field, in an extremely thorough way. Training is offered in Ontario, Canada, where thousands of farms have completed this process. The province offers cost sharing and tax credits for farms that have completed the evaluation and had their plan approved by a peer-review panel.

Farm-A-Syst

Extension Agricultural Engineering
237 Seaton Hall
Kansas State University
Manhattan, KS 66506
Attn: Danny Rogers
Phone: (785) 532-5813

The focus of this tool is on farmstead water quality risks. NRCS Farm Evaluation sheets cover pasture, cropland, and woodlands, field by field rather than whole farm. See www.wisc.edu/farmasyst/Planetor for a computerized environmental assessment tool, field by field.

New York City Watershed and Skaneateles Watershed

These projects are funded by the cities that rely on these watersheds for drinking water. After a lengthy political process, the people in these watersheds, rather than building expensive filtering systems, decided to provide funds for farms in the watershed to assess where they might be polluting, design solutions to eliminate the pollution, and then make the changes necessary. Elizabeth Henderson found that the farmers who went through this process felt that it gave a great lift to their farming, helping them eliminate old problems and giving them lots of new ideas and enthusiasm for their farming.

Janke, Rhonda, and Stan Freyenburger, *Indicators of Sustainability in Whole-Farm Planning: Planning Tools*, Kansas Sustainable Agriculture Series, Paper #3; www.oznet.ksu.edu. This paper lists many other planning tools and provides brief descriptions.

Index

About the
Authors and Illustrator

Elizabeth Henderson has been growing organic fruits and vegetables for the fresh market for over thirty years. From 1998 through 2010, she farmed with her partners, Ammie Chickering and Greg Palmer, at Peacework Organic Farm in Wayne County, New York, producing shares for Genesee Valley Organic Community Supported Agriculture, in its 23th year in 2011, and has retired from full-time farming. She is one of the authors of *The Real Dirt: Farmers Tell About Organic and Low-Input Agriculture in the Northeast, Food Book for Sustainable Harvest,* and *Sharing the Harvest: A Citizen's Guide to Community Supported Agriculture.*

Karl North, after his graduate work in social anthropology and ecology, learned and practiced subsistence agriculture in the French Pyrenees. With his wife, Jane, he built and has operated Northland Sheep Dairy since 1980. He completed three years of training as an educator in holistic management in 2004. His articles on the themes of holism and sustainability in agriculture have appeared in NOFA periodicals.

Illustrator Jocelyn Langer is an artist, music teacher, and organic gardener, and the illustrator of the NOFA organic farming handbooks. She illustrates and does graphic design work for alternative media and political events as well as organic-farming-related publications. Jocelyn lives in central Massachusetts.

The special farmer-reviewers for this manual were Doug and Sarah Flack; the scientific reviewer was Ed Martsolf.

"This logo identifies paper that meets the standards of the Forest Stewardship Council. FSC is widely regarded as the best practice in forest management, ensuring the highest protections for forests and indigenous peoples."